COMPREHENSIVE FOOD FERMENTATION AND BIOTECHNOLOGY

COMPREHENSIVE FOOD FERMENTATION AND BIOTECHNOLOGY

Ahtesham Malik

RANDOM PUBLICATIONS
NEW DELHI - 110 002 (INDIA)

Comprehensive Food Fermentation and Biotechnology

ISBN 978-93-51116-95-0

Published in 2015 in India by

RANDOM PUBLICATIONS

4376-A/4B, Gali Murari Lal, Ansari Road
New Delhi-110 002
Phone: +9111-43580356, 23289044
E-mail: randomexports@gmail.com; sales@randompublications.com; info@randompublications.com

Reprinted 2026

Type Setting by: Friends Media, Delhi-110089

Preface

Fermentation is a metabolic process that converts sugar to acids, gases and/or alcohol. It occurs in yeast and bacteria, but also in oxygen-starved muscle cells, as in the case of lactic acid fermentation. Fermentation is also used more broadly to refer to the bulk growth of microorganisms on a growth medium. French microbiologist Louis Pasteur is often remembered for his insights into fermentation and its microbial causes. The science of fermentation is known as zymology. Fermentation takes place in the lack of oxygen (when the electron transport chain is unusable) and becomes the cell's primary means of ATP (energy) production. It turns NADH and pyruvate produced in the glycolysis step into NAD^+ and various small molecules depending on the type of fermentation. In the presence of O_2, NADH and pyruvate are used to generate ATP in respiration. This is called oxidative phosphorylation, and it generates much more ATP than glycolysis alone. For that reason, cells generally benefit from avoiding fermentation when oxygen is available. Exceptions include obligate anaerobes, which cannot tolerate oxygen. In oxidative phosphorylation the energy for ATP formation is derived from an electrochemical proton gradient generated across the inner mitochondrial membrane (or, in the case of bacteria, the plasma membrane) via the electron transport chain. Glycolysis has substrate-level phosphorylation (ATP generated directly at the point of reaction).

Fermentation has been used by humans for the production of food and beverages since the Neolithic age. For example, fermentation is employed for preservation in a process that produces lactic acid as found in such sour foods as pickled cucumbers, kimchi and yogurt, as well as for producing alcoholic beverages such as wine and beer. Fermentation can even occur within the stomachs of animals, such as humans. Auto-brewery syndrome is a rare medical condition where the stomach contains brewers yeast that break down starches into ethanol; which enters the blood stream. Sugars are the most common substrate of fermentation, and typical examples of fermentation products are ethanol, lactic acid, carbon dioxide, and hydrogen gas (H_2). However, more exotic compounds can be produced by fermentation, such as butyric acid and acetone. Yeast carries out fermentation in the production of ethanol in beers, wines, and other alcoholic drinks, along with the production of large quantities of carbon dioxide. Fermentation occurs in mammalian muscle during periods of intense exercise where oxygen supply becomes limited, resulting in the creation of lactic acid. Humans have taken advantage of the metabolism in a tiny fungus called yeast to create beer and wine from grains and fruits. Advances in microbiology and fermentation technology have continued steadily up until the present. For example, in the late 1970s, it was discovered that microorganisms could be mutated with physical and chemical treatments to be higher-yielding, faster-growing, tolerant of less oxygen, and able to use a more concentrated medium. Strain selection and hybridization developed as well, affecting most modern food fermentations. Biotechnology has a major application in the food sector. It helps in improving the edibility, texture, and storage of the food; in preventing the attack of the food, mainly dairy, by the virus like bacteriophage; producing antimicrobial effect to destroy the unwanted microorganisms in food that cause toxicity; to prevent the formation of mycotoxins; and degradation of other toxins and anti-nutritional elements present naturally in food. The advance in application of biotechnology in food processing mainly concerns with the traditional approach of improving the strains of microorganisms and development of further improved micro-organic derivatives like enzymes, etc.

This book will serve as a valuable resource for students, food scientists, technologists, microbiologists, toxicologists, and processors.

I thank all members of my team who have helped in the preparation of the book. My special thanks go to "Random Publications" who have published the book.

— Ahtesham Malik

Contents

Chapter 1

Food Contaminants

Food contamination refers to the presence in food of harmful chemicals and microorganisms which can cause consumer illness. This article addresses the chemical contamination of foods, as opposed to microbiological contamination, which can be found under Foodborne illness. A separate issue isgenetically modified food, or the presence in foods of ingredients from genetically modified organisms, also referred to as a form of food contamination.

Impact

The impact of chemical contaminants on consumer health and well-being is often apparent only after many years of prolonged exposure at low levels (e.g. cancer). Chemical contaminants present in foods are often unaffected by thermal processing (unlike most microbiological agents). Chemical contaminants can be classified according to the source of contamination and the mechanism by which they enter the food product.

Agrochemicals

Agrochemicals are chemicals used in agricultural practices and animal husbandry with the intent to increase crops and reduce costs. Such agents includepesticides (e.g. insecticides, herbicides, rodenticides), plant growth regulators, veterinary drugs (e.g. nitrofuran, fluoroquinolones, malachite green, chloramphenicol), and bovine somatotropin (rBST).

Environmental Contaminants

Environmental contaminants are chemicals that are present in the environment in which the food is grown, harvested, transported, stored, packaged, processed, and consumed. The physical contact of the food with its environment results in its contamination. Possible sources of contamination are:

Air: radionuclides (Caesium, Strontium), polycyclic aromatic hydrocarbons (PAH).

Water: arsenic, mercury.

Soil: cadmium, nitrates, perchlorates.

Polychlorinated biphenyls (PCB), dioxins, and polybrominated diphenyl ethers (PBDE) are ubiquitous chemicals, which are present in air, water, soil, and the entire biosphere.

Packaging materials: antimony, tin, lead, perfluorooctanoic acid (PFOA), semicarbazide, benzophenone, isopropylthioxanthone (ITX), bisphenol A.

Processing/cooking equipment: copper, or other metal chips, lubricants, cleaning and sanitizing agents.

Naturally occurring toxins: mycotoxins, phytohaemagglutinin, pyrrolizidine alkaloids, grayanotoxin, mushroom toxins, scombrotoxin (histamine), ciguatera, shellfish toxins (see shellfish poisoning), tetrodotoxin, among many others.

Banned Pesticides and Carcinogens

There are many cases of banned pesticides or carcinogens found in foods.

- Greenpeace exposed in 2006 in China that 25% of surveyed supermarkets agricultural products contained banned pesticides. Over 70% of tomatoes that tested were found to have the banned pesticide Lindane, and almost 40% of the samples had a mix of three or more types of pesticides. Fruits were also tested in this investigation. Tangerines, strawberries and Kyofung grapes samples were found contaminated by banned pesticides, including the highly toxic Methamidophos. These fruits can also be found in Hong Kong market. Greenpeace says there exists no comprehensive monitoring on fruit produce in the Hong Kong as of 2006.
- In India, soft drinks were found contaminated with high levels of pesticides and insecticides, including lindane, DDT, malathion and chlorpyrifos.
- News of Formaldehyde, a carcinogen was found in Vietnamese national dish, Pho, broke in 2007 Vietnam food scare. Vegetables and fruits were also found to have banned pesticides. "Health

agencies have known that Vietnamese soy sauce, the country's second most popular sauce after fish sauce, has been chock full of cancer agents since at least 2001," thundered the Thanh Niendaily. "Why didn't anyone tell us?" The carcinogen in Asian sauces is 3-MCPD and its metabolite 1,3-DCP, which has been an ongoing problem before 2000 affecting multiple continents.

- 2005 Indonesia food scare, carcinogenic formaldehyde was added as a preservative to noodles, tofu, salted fish, and meatballs.
- 2008 Chinese milk scandal.

Hair in Food

There is a heavy stigma attached to the presence of hair in food in most societies. There is a risk that it may induce choking and vomiting, and also that it may be contaminated by toxic substances. Views differ as to the level of risk it poses to the inadvertent consumer.

In most countries, people working in the food industry are required to cover their hair. When people are served food which contains hair in restaurants or cafés, it is usual for them to complain to the staff. Despite this, it is not a valid ground on which to sue the restaurant in the United States but in the United Kingdom it breaks the regulations of the UK Food Safety Act 1990 as it has been known to cause food poisoning and people can sue over this.

There are a range of possible reasons for the objection to hair in food, ranging from cultural taboos to the simple fact that it is difficult to digest and unpleasant to eat. It may also be interpreted as a sign of more widespread problems with hygiene. The introduction of complete capture hairnets is believed to have resulted in a decrease in incidents of contamination of this type.

Sometimes protein from human hair is used as a food ingredient, in bread and other such similar products. Such use of human hair in food is forbidden in Islam. Historically in Judaism, finding hair in food was a sign of bad luck.

Processing Contaminants

Processing contaminants are generated during the processing of foods (e.g. heating, fermentation). They are absent in the raw materials, and are formed by chemical reactions between natural and/or added food constituents during processing. The presence of these contaminants in processed foods cannot be entirely avoided. Technological processes can be adjusted and/or optimized, however, in order to reduce the levels of formation of processing contaminants.

Examples are: nitrosamines, polycyclic aromatic hydrocarbons (PAH), heterocyclic amines, histamine, acrylamide, furan, benzene,trans fat, monochloropropanediol (MCPD), semicarbazide, 4-hydroxynonenal (4-HNE), and ethyl carbamate. There is also the possibility of metal chips from the processing equipment contaminating food. These can be identified using metal detection equipment. In many conveyor lines, the line will be stopped, or when weighing the product with a Check weigher, the item can be rejected for being over-or underweight or because small pieces of metal are detected within it.

Emerging Food Contaminants

While many food contaminants have been known for decades, the formation and presence of certain chemicals in foods has been discovered relatively recently. These are the so-called emerging food contaminants, e.g. acrylamide, furan, benzene, perchlorate, perfluorooctanoic acid (PFOA), 3-monochloropropane-1,3-diol (3-MCPD), 4-hydroxynonenal and (4-HNE).

Safety and Regulation

Acceptable Daily Intake (ADI) levels and tolerable concentrations of contaminants in individual foods are determined on the basis of the "No Observed Adverse Effect Level" (NOAEL) in animal experiments, by using a safety factor (usually 100). The maximum concentrations of contaminants allowed by legislation are often well below toxicological tolerance levels, because such levels can often be reasonably achieved by using good agricultural and manufacturing practices.

The establishment of ADIs for certain emerging food contaminants is currently an active area of research and regulatory debate.

Food Safety

Food safety is a scientific discipline describing handling, preparation, and storage of food in ways that prevent foodborne illness. This includes a number of routines that should be followed to avoid potentially severe health hazards. Food can transmit disease from person to person as well as serve as a growth medium for bacteria that can cause food poisoning.

Debates on genetic food safety include such issues as impact of genetically modified food on health of further generations and genetic pollution of environment, which can destroy natural biological diversity. In developed countries there are intricate standards for food preparation, whereas in lesser developed countries the main issue is simply the availability of adequate safe water, which is usually a critical item. In theory food poisoning is 100% preventable.

Key Principles

The five key principles of food hygiene, according to WHO, are:

1. Prevent contaminating food with pathogens spreading from people, pets, and pests.
2. Separate raw and cooked foods to prevent contaminating the cooked foods.
3. Cook foods for the appropriate length of time and at the appropriate temperature to kill pathogens.
4. Store food at the proper temperature.
5. Use safe water and raw materials.

ISO 22000

ISO 22000 is a standard developed by the International Organization for Standardization dealing with food safety. This is a general derivative of ISO 9000. ISO 22000 standard: The ISO 22000 international standard specifies the requirements for a food safety management system that involves interactive communication, system management, prerequisite programs, HACCPprinciples.

Incidence

A 2003 World Health Organization (WHO) report concluded that about 40% of reported food poisoning outbreaks in the WHO European Region occur in private homes. According to the WHO and CDC, in the USA alone, annually, there are 76 million cases of foodborne illness leading to 325,000 hospitalizations and 5,000 deaths.

Regulatory Agencies

Australia

Australian Food Authority is working toward ensuring that all food businesses implement food safety systems to ensure food is safe to consume in a bid to halt the increasing incidence of food poisoning, this includes basic food safety training for at least one person in each business. Smart business operators know that basic food safety training improves the bottom line, staff take more pride in their work; there is less waste; and customers can have more confidence in the food they consume.

Food Safety training in units of competence from a relevant training package, must be delivered by a Registered Training Organization

(RTO) to enable staff to be issued with a nationally-recognised unit of competency code on their certificate. Generally this training can be completed in less than one day. Training options are available to suit the needs of everyone. Training may be carried out in-house for a group, in a public class, via correspondence or online. Basic food safety training includes:

- Understanding the hazards associated with the main types of food and the conditions to prevent the growth of bacteria which can cause food poisoning
- The problems associated with product packaging such as leaks in vacuum packs, damage to packaging or pest infestation, as well as problems and diseases spread by pests.
- Safe food handling. This includes safe procedures for each process such as receiving, re-packing, food storage, preparation and cooking, cooling and re-heating, displaying products, handling products when serving customers, packaging, cleaning and sanitizing, pest control, transport and delivery. Also the causes of cross contamination.
- Catering for customers who are particularly at risk of food-borne illness, including allergies and intolerance.
- Correct cleaning and sanitizing procedures, cleaning products and their correct use, and the storage of cleaning items such as brushes, mops and cloths.
- Personal hygiene, hand washing, illness, and protective clothing.

People responsible for serving unsafe food can be liable for heavy fines under this new legislation, consumers are pleased that industry will be forced to take food safety seriously.

China

Food safety is a growing concern in Chinese agriculture. The Chinese government oversees agricultural production as well as the manufacture of food packaging, containers, chemical additives, drug production, and business regulation. In recent years, the Chinese government attempted to consolidate food regulation with the creation of the State Food and Drug Administration in 2003, and officials have also been under increasing public and international pressure to solve food safety problems. However, it appears that regulations are not well known by the trade. Labels used for "green" food, "organic" food and "pollution-free" food are not well recognized by traders and many are unclear about their meaning. A survey by the World Bank found that supermarket managers had difficulty in obtaining produce that met safety requirements and found that a high percentage of produce did not comply with established standards.

Traditional marketing systems, whether in China or the rest of Asia, presently provide little motivation or incentive for individual farmers to make improvements to either quality or safety as their produce tends to get grouped together with standard products as it progresses through the marketing channel. Direct linkages between farmer groups and traders or ultimate buyers, such as supermarkets, can help avoid this problem. Governments need to improve the condition of many markets through upgrading management and reinvesting market fees in physical infrastructure. Wholesale markets need to investigate the feasibility of developing separate sections to handle fruits and vegetables that meet defined safety and quality standards.

European Union

The parliament of the European Union (EU) makes legislation in the form of directives and regulations, many of which are mandatory for member states and which therefore must be incorporated into individual countries' national legislation. As a very large organisation that exists to remove barriers to trade between member states, and into which individual member states have only a proportional influence, the outcome is often seen as an excessively bureaucratic 'one size fits all' approach. However, in relation to food safety the tendency to err on the side of maximum protection for the consumer may be seen as a positive benefit. The EU parliament is informed on food safety matters by the European Food Safety Authority.

Individual member states may also have other legislation and controls in respect of food safety, provided that they do not prevent trade with other states, and can differ considerably in their internal structures and approaches to the regulatory control of food safety.

Germany

The Federal Ministry of Food, Agriculture and Consumer Protection (BMELV) is a Federal Ministry of the Federal Republic of Germany. History: Founded as Federal Ministry of Food, Agriculture and Foresting in 1949, this name did not change until 2001. Then the name changed to Federal Ministry of Consumer Protection, Food and Agriculture. At the 22nd of November 2005, the name got changed again to its current state: Federal Ministry of Food, Agriculture and Consumer Protection. The reason for this last change was that all the resorts should get equal ranking which was achieved by sorting the resorts alphabetically. Vision: A balanced and healthy diet with safe food, distinct consumer rights and consumer information for various areas of life, and a strong and sustainable agriculture as well as perspectives for our rural areas are important goals of the Federal Ministry of Food, Agriculture and Consumer Protection (BMELV). The Federal Office of Consumer Protection and Food Safety is under the control of the Federal Ministry of Food, Agriculture and Consumer Protection. It exercises several duties, with which it contributes to safer food and thereby intensifies health-based consumer protection in Germany. Food can be manufactured and sold within Germany without a special permission, as long as it does not cause any damage on consumers' health and meets the general standards set by the legislation. However, manufacturers, carriers, importers and retailers are responsible for the food they pass into circulation. They are obliged to ensure and document the safety and quality of their food with the use of in-house control mechanisms.

Hong Kong

In Hong Kong SAR, the Centre for Food Safety is in charge of ensuring food sold is safe and fit for consumption.

Pakistan

Pakistan does not have an integrated legal framework but has a set of laws, which deals with various aspects of food safety. These laws, despite the fact that they were enacted long time ago, have tremendous capacity to achieve at least minimum level of food safety. However, like many other laws, these laws remain very poorly enforced. There are four laws that specifically deal with food safety. Three of these laws directly focus issues related to food safety, while the fourth one namely Pakistan Standards and Quality Control Authority Act, is indirectly relevant to food safety. The Pure Food Ordinance, 1960 consolidates and amends the law in relation to the preparation and the sale of foods. All provinces and some northern areas have adopted this law with certain amendments. Its aim is to ensure purity of food being supplied to people in the market and, therefore, provides for preventing adulteration. The Pure Food Ordinance 1960 does not apply to cantonment areas. There is separate law for cantonments called "The Cantonment Pure Food Act, 1966". There is no substantial difference between the Pure Food Ordinance 1960 and The Cantonment Pure Food Act. Even the rules of operation are very much similar. Pakistan Hotels and Restaurant Act, 1976 applies to all hotels and restaurants in Pakistan and seeks to control and regulate the rates and standard of service(s) by hotels and restaurants. In addition to other provisions, under section 22(2), the sale of food or beverages that are contaminated, not prepared hygienically or served in utensils that are not hygienic or clean is an offence. There are no express provisions for consumer complaints in the Pakistan Restaurants Act, 1976, Pakistan Penal Code, 1860 and Pakistan Standards and Quality Control Authority Act, 1996. However, the laws do not prevent citizens from lodging complaints with the concerned government officials. However, the consideration and handling of complaints is a matter of discretion of the officials.

South Korea

Korea Food and Drug Administration

Korea Food & Drug Administration (KFDA) is working for food safety since 1945. It is part of the Government of South Korea.

IOAS-Organic Certification Bodies Registered in KFDA: "Organic" or related claims can be labelled on food products when organic certificates are considered as valid by KFDA. KFDA admits organic certificates which can be issued by 1) IFOAM (International Federation of Organic Agriculture Movement) accredited certification bodies 2) Government accredited certification bodies-328 bodies in 29 countries have been registered in KFDA.

Food Import Report: According to Food Import Report, it is supposed to report or register what you import. Competent authority is as followed:

- Imported Agricultural Products, Processed Foods, Food Additives, Utensils, Containers & Packages or Health Functional Foods
  - — KFDA (Korea Food and Drug Administration)
- Imported Livestock, Livestock products (including Dairy products)
  - — NVRQS (National Veterinary Research and Quarantine Service)-Packaged meat, milk & dairy products (butter, cheese), hamburger patties, meat ball and other processed products which are stipulated by Livestock Sanitation Management Act
- Imported Marine products
  - — NFIS (National Fisheries Products Quality Inspection Service)-Fresh, chilled, frozen, salted, dehydrated, eviscerated marine produce which can be recognized its characteristics.

National Institute of Food and Drug Safety Evaluation

National Institute of Food and Drug Safety Evaluation (NIFDS) is functioning as well. The National Institute of Food and Drug Safety Evaluation is a national organization for toxicological tests and research. Under the Korea Food & Drug Administration, the Institute performs research on toxicology, pharmacology, and risk analysis of foods, drugs, and their additives. The Institute strives primarily to understand important biological triggering mechanisms and improve assessment methods of human exposure, sensitivities, and risk by (1) conducting basic, applied, and policy research that closely examines biologically triggering harmful effects on the regulated products such as foods, food additives, and drugs, and (2) operating the national toxicology program for the toxicological test development and inspection of hazardous chemical ubstances assessments. The Institute ensures safety by (1) investigation and research on safety by its own researchers, (2) contract research by external academicians and research centres.

United States

The US food system is regulated by numerous federal, state and local officials. Although the US food safety system is one of the best in the world, it is lacking in "organization, regulatory tools, and resources to address food borne illness."

Federal Level Regulation

The Food and Drug Administration publishes the Food Code, a model set of guidelines and procedures that assists food control jurisdictions by providing a scientifically sound technical and legal basis for regulating the retail and food service industries, including restaurants, grocery stores and institutional food service providers such as nursing homes.

Regulatory agencies at all levels of government in the United States use the FDA Food Code to develop or update food safety rules in their jurisdictions that are consistent with national food regulatory policy. According to the FDA, 48 of 56 states and territories, representing 79% of the U.S. population, have adopted food codes patterned after one of the five versions of the Food Code, beginning with the 1993 edition.

In the United States, federal regulations governing food safety are fragmented and complicated, according to a February 2007 report from the Government Accountability Office. There are 15 agencies sharing oversight responsibilities in the food safety system, although the two primary agencies are the U.S. Department of Agriculture

(USDA) Food Safety and Inspection Service (FSIS), which is responsible for the safety of meat, poultry, and processed egg products, and the Food and Drug Administration (FDA), which is responsible for virtually all other foods.

The Food Safety and Inspection Service has approximately 7,800 inspection program personnel working in nearly 6,200 federally inspected meat, poultry and processed egg establishments. FSIS is charged with administering and enforcing the Federal Meat Inspection Act, the Poultry Products Inspection Act, the Egg Products Inspection Act, portions of the Agricultural Marketing Act, the Humane Slaughter Act, and the regulations that implement these laws. FSIS inspection program personnel inspect every animal before slaughter, and each carcass after slaughter to ensure public health requirements are met. In fiscal year (FY) 2008, this included about 50 billion pounds of livestock carcasses, about 59 billion pounds of poultry carcasses, and about 4.3 billion pounds of processed egg products. At U.S. borders, they also inspected 3.3 billion pounds of imported meat and poultry products.

Industry Pressure

There have been concerns over the efficacy of safety practices and food industry pressure on U.S. regulators. A study reported by Reuters found that "the food industry is jeopardizing U.S. public health by withholding information from food safety investigators or pressuring regulators to withdraw or alter policy designed to protect consumers". A survey found that 25% of U.S. government inspectors and scientists surveyed have experienced during the past year corporate interests forcing their food safety agency to withdraw or to modify agency policy or action that protects consumers. Scientists have observed that management undercuts field inspectors who stand up for food safety against industry pressure. According to Dr. Dean Wyatt, a USDA veterinarian who oversees federal slaughter house inspectors, "Upper level management does not adequately support field inspectors and the actions they take to protect the food supply. Not only is there lack of support, but there's outright obstruction, retaliation and abuse of power."

State and Local Regulation

A number of U.S. states have their own meat inspection programs that substitute for USDA inspection for meats that are sold only in-state. Certain state programs have been criticized for undue leniency to bad practices. However, other state food safety programs supplement, rather than replace, Federal inspections, generally with the goal of increasing consumer confidence in the state's produce. For example, state health departments have a role in investigating outbreaks of food-borne disease bacteria, as in the case of the 2006 outbreak of *Escherichia coli* O157:H7 (bad *E. coli* bacteria) from processed spinach. Health departments also promote better food processing practices to eliminate these threats.

In addition to the US Food and Drug Administration, several states that are major producers of fresh fruits and vegetables (including California, Arizona and Florida) have their own state programs to test produce for pesticide residues.

Restaurants and other retail food establishments fall under state law and are regulated by state or local health departments. Typically these regulations require official inspections of specific design features, best food-handling practices, and certification of food handlers. In some places a letter grade or numerical score must be prominently posted following each inspection. In some localities, inspection deficiencies and remedial action are posted on the Internet.

Manufacturing Control

HACCP Guidelines

The UK Food Standards Agency publishes recommendations as part of its Hazard Analysis and Critical Control Points (HACCP) programme. The relevant guidelines state that:

"Cooking food until the CORE TEMPERATURE is 75 °C or above will ensure that harmful bacteria are destroyed.

However, lower cooking temperatures are acceptable provided that the CORE TEMPERATURE is maintained for a specified period of time as follows :

- 60 °C for a minimum of 45 minutes
- 65 °C for a minimum of 10 minutes
- 70 °C for a minimum of 2 minutes".

Previous guidance from a leaflet produced by the UK Department Of Health "Handling Cooked Meats Safely A Ten Point Plan" also allowed for:

- "75 °C for a minimum of 30 seconds
- 80 °C for a minimum of 6 seconds".

Note that recommended cooking conditions are only appropriate if initial bacterial numbers in the uncooked food are small. Cooking does not overcome poor hygiene.

Consumer Labelling

United Kingdom

Food stuffs in the UK have one of two labels to indicate the nature of the deterioration of the product and any subsequent health issues. EHO Food Hygiene certification is required to prepare and distribute food. While there is no specified expiry date of such a qualification the changes in legislation it is suggested to update every five years.

Best before indicates a future date beyond which the food product *may* lose quality in terms of taste or texture amongst others, but does not imply any serious health problems if food is consumed beyond this date (within reasonable limits).

Use by indicates a legal date beyond which it is not permissible to sell a food product (usually one that deteriorates fairly rapidly after production) due to the potential serious nature of consumption of pathogens. Leeway is sometimes provided by producers in stating display until dates so that products are not at their limit of safe consumption on the actual date stated (this latter is voluntary and not subject to regulatory control). This allows for the variability in production, storage and display methods.

United States

With the exception of infant formula and baby foods which must be withdrawn by their expiration date, Federal law does not require expiration dates. For all other foods, except dairy products in some states, freshness dating is strictly voluntary on the part of manufacturers. In response to consumer demand, perishable foods are typically labeled with a Sell by date. It is up to the consumer to decide how long after the Sell by date a package is usable. Other common dating statements are Best if used by, Use-by date, Expiration date, Guaranteed fresh <date>, and Pack date.

Australia and New Zealand

Guide to Food Labelling and Other Information Requirements: This guide provides background information on the general labelling requirements in the Code. The information in this guide applies both to food for retail sale and to food for catering purposes. Foods for catering purposes means those foods for use in restaurants, canteens, schools, caterers or self-catering institutions, where food is offered for immediate consumption. Labelling and information requirements in the new Code apply both to food sold or prepared for sale in Australia and New Zealand and food imported into Australia and New Zealand. Warning and Advisory Declarations, Ingredient Labelling, Date Marking, Nutrition Information Requirements, Legibility Requirements for Food Labels, Percentage Labelling, Information Requirements for Foods Exempt from Bearing a Label.

Issues Associated with Sell by/use by Dates

According to the UK's Waste & Resources Action Programme, 33% percent of all food produced is wasted along the chill chain or at the consumer. At the same time, a large number of people get sick every year due to spoiled food.

UK Government to replace Sell by/use by Dates?

According to the UK minister Hilary Benn the use by date and sell by dates are old technologies that are outdated and should be replaced by other solutions or disposed of altogether.

How to Enhance Food Safety

There is a number of ways to enhance sell by and use by dates. These include better education of consumers on how to use, transport, and store fresh food products, but also by enhancing the use by and sell by dates by adding to the package smart indicators such as TTIs (Time Temperature Indicators). These show through a visible colour change whether the product is still fresh. TTIs are already in use by retailers and food producers in France (Monoprix and Carrefour), Switzerland (Kneuss), and other countries in western Europe.

Consumer Tips: How to Keep Food Safe

1. One of the simplest measures that any person can take to prevent the spread of foodborne illness is to properly wash his or her hands before preparing or eating any meal. Many people who believe they are adequately washing their hands are sorely mistaken. According to a study conducted by the American Society of Microbiology, 97% of females and 92% of males said they washed their hands, but those numbers turned out to be 75% of females and 58% of males upon observation. A proper hand-washing technique suggested by the Missouri Department of Health and Senior Services includes using soap and warm water; washing vigorously for 10-20 seconds, making sure to reach all surfaces of the hands including the wrists, between the fingers, and under the finger nails; rinsing well; drying hands with a paper towel; and using a paper towel to turn off the water. It is certainly important to wash your hands before preparing a meal and eating, but hands should also be washed

after using the restroom, coughing/sneezing, touching cuts or skin infections, handling raw meat, and touching pets or other animals. Almost half the cases of foodborne illnesses could be prevented by better hand-washing by food handlers. In addition, hand sanitizer is a helpful follow-up to hand-washing, but it should never be used to replace this valuable technique.

2. Some foods should simply never be ingested in the first place because they have such a high risk of containing harmful bacteria that can make people ill. These foods include raw milk, unpasteurized dairy products, unpasteurized juices (such as fresh apple cider), raw meat, and raw cookie dough.
3. In cases of suspected food spoilage, food should never simply be tasted, smelled, or eye-balled in order to determine its safety. It is true that in some cases, the presence of mold or other growths may indicate that a food has reached its expiration date and should not be consumed. However, some foods may look and smell fine, but if they have been stored at room temperature for over 2 hours, microscopic bacteria may have been allowed to grow. The safe temperature for foods in refrigerators is between 35°F and 45°F, and freezers should be kept at 0° F or below.
4. It is not safe to let meats thaw on the counter all day, because this allows any germs present on the food to thrive. Safer alternatives to this practice include thawing the food under running water (70°F or below) for less than 2 hours, placing the food in the refrigerator to thaw, or thawing the food in the microwave as part of the cooking process.
5. It is also important to make sure that raw meats are cooked to the appropriate internal temperature before they are consumed. Safe internal temperatures for various meats include 165°F for poultry, 155°F for ground meat, and 145°F for fish and pork. It is not sufficient to judge the doneness of meat by its internal colour. According to a study performed by the USDA, –25% of hamburgers with a brown internal colour were not cooked to the proper temperature." Instead of simply eye balling the meat, it is essential to use a meat thermometer to judge the safety of consuming the food.
6. The cooking process is often the time that foodborne pathogens are allowed to enter the food we eat because of the prevalence of cross contamination. Cross contamination occurs when a person handling raw meats, eggs, fish, or other foods containing

harmful pathogens touches cooking utensils, cutting boards, or cooking surfaces and spreads the pathogens to ready-to-eat foods in the process. This mode of transmission can be interrupted by washing hands after handling raw foods, washing utensils and cutting boards that have come in contact with raw foods, and disinfecting counter surfaces frequently.

7. Leftovers are ideally stored in the refrigerator in shallow containers (2 inches tall or less) so that the cooling process can be accelerated and the buildup of harmful bacteria can be prevented. Storing foods in larger containers may keep foods warm and allow harmful bacteria to grow.

Food Preservation

Food preservation is the process of treating and handling food to stop or slow down spoilage (loss of quality, edibility or nutritional value).

Preservation usually involves preventing the growth of bacteria, yeasts, fungi, and other micro-organisms (although some methods work by introducing benignbacteria, or fungi to the food), as well as retarding the oxidation of fats which cause rancidity. Black preservation also includes processes which inhibit natural discoloration that can occur during food preparation, such as the enzymatic browning reaction in apples after they are cut.

Many processes designed to preserve food will involve a number of food preservation methods. Preserving fruit, by turning it into jam, for example, involves boiling (to reduce the fruit's moisture content and to kill bacteria, yeasts, etc.), sugaring (to prevent their re-growth) and sealing within an airtight jar (to prevent recontamination). There are many traditional methods of preserving food that limit the energy inputs and reduce carbon footprint. Maintaining or creating nutritional value, texture and flavour is an important aspect of food preservation, although, historically, some methods drastically altered the character of the food being preserved. In many cases these changes have now come to be seen as desirable qualities – cheese, yoghurt and pickled onions being common examples.

Preservation Processes

Preservation processes include:

- Heating to kill or denature micro-organisms (e.g. boiling)
- Oxidation (e.g. use of sulphur dioxide)
- Toxic inhibition (e.g. smoking, use of carbon dioxide, vinegar, alcohol etc.)

- Dehydration (drying)
- Osmotic inhibition (e.g. use of syrups)
- Low temperature inactivation (e.g. freezing)
- Ultra high water pressure (e.g. fresherized, a kind of "cold" pasteurization, the pressure kills naturally occurring pathogens, which cause food deterioration and affect food safety.)
- Combinations of these methods.

Drying

One of the oldest methods of food preservation is by drying, which reduces water activity sufficiently to prevent or delay bacterial growth.

Refrigeration

Refrigeration preserves food by slowing down the growth and reproduction of micro-organisms and the action of enzymes which cause food to rot. The introduction of commercial and domestic refrigerators drastically improved the diets of many in the Western world by allowing foods such as fresh fruit, salads and dairy products to be stored safely for longer periods, particularly during warm weather.

Freezing

Freezing is also one of the most commonly used processes commercially and domestically for preserving a very wide range of food including prepared food stuffs which would not have required freezing in their unprepared state. For example, potato waffles are stored in the freezer, but potatoes themselves require only a cool dark place to ensure many months' storage. Cold stores provide large volume, long-term storage for strategic food stocks held in case of national emergency in many countries.

Heat Treating

Vacuum Packing

Vacuum-packing stores food in a vacuum environment, usually in an air-tight bag or bottle. The vacuum environment strips bacteria of oxygen needed for survival, slowing spoiling. Vacuum-packing is commonly used for storing nuts to reduce loss of flavour from oxidation. Chips are vacuum packed.

Salt

Salting or curing draws moisture from the meat through a process of osmosis. Meat is cured with salt or sugar, or a combination of the two. Nitrates and nitrites are also often used to cure meat and contribute the characteristic pink colour, as well as inhibition of Clostridium botulinum.

Sugar

Sugar is used to preserve fruits, either in syrup with fruit such as apples, pears, peaches, apricots, plums or in crystallized form where the preserved material is cooked in sugar to the point of crystallisation and the resultant product is then stored dry. This method is used for the skins of citrus fruit (candied peel), angelica and ginger. A modification of this process produces glacé fruit such as glacé cherries where the fruit is preserved in sugar but is then extracted from the syrup and sold, the preservation being maintained by the sugar content of the fruit and the superficial coating of syrup. The use of sugar is often combined with alcohol for preservation of luxury products such as fruit in brandy or other spirits. These should not be confused with fruit flavoured spirits such as cherry brandy or Sloe gin.

Artificial Food Additives

Preservative food additives can be *antimicrobial*; which inhibit the growth of bacteria or fungi, including mold, or *antioxidant*; such as oxygen absorbers, which inhibit the oxidation of food constituents. Common antimicrobial preservatives include calcium propionate, sodium nitrate, sodium nitrite, sulfites (sulfur dioxide, sodium bisulfite, potassium hydrogen sulfite, etc.) and disodium EDTA. Antioxidants include BHA and BHT. Other preservatives include formaldehyde (usually in solution), glutaraldehyde (kills insects), ethanol and methylchloroisothiazolinone.

Pickling

Pickling is a method of preserving food in an edible anti-microbial liquid. Pickling can be broadly categorized as chemical pickling for example, In chemical pickling, the food is placed in an edible liquid that inhibits or kills bacteria and other micro-organisms. Typical pickling agents include brine (high in salt), vinegar, alcohol, and vegetable oil, especially olive oil but also many other oils. Many chemical pickling processes also involve heating or boiling so that the food being preserved becomes saturated with the pickling agent. Common chemically pickled foods include cucumbers, peppers, corned beef, herring, and eggs, as well mixed vegetables such as piccalilli.

In fermentation pickling, the food itself produces the preservation agent, typically by a process that produces lactic acid. Fermented pickles include sauerkraut, nukazuke, kimchi, surströmming, and curtido. Some pickled cucumbers are also fermented.

In commercial pickles, a preservative like sodium benzoate or EDTA may also be added to enhance shelf life. You can also say adding salt and oil is also pickling.

Lye

Sodium hydroxide (lye) makes food too alkaline for bacterial growth. Lye will saponify fats in the food, which will change its flavour and texture. Lutefisk uses lye in its preparation, as do some olive recipes. Modern recipes for century eggs also call for lye. Masa harina and hominy use agricultural lime in their preparation and this is often misheard as 'lye'.

Canning and Bottling

Canning involves cooking food, sealing it in sterile cans or jars, and boiling the containers to kill or weaken any remaining bacteria as a form of sterilization. It was invented by Nicolas Appert . Foods have varying degrees of natural protection against spoilage and may require that the final step occur in a pressure cooker. High-acid fruits like strawberries require no preservatives to can and only a short boiling cycle, whereas marginal fruits such as tomatoes require longer boiling and addition of other acidic elements. Low acid foods, such as vegetables and meats require pressure canning. Food preserved by canning or bottling is at immediate risk of spoilage once the can or bottle has been opened. Lack of quality control in the canning process may allow ingress of water or micro-organisms. Most such failures are rapidly detected as decomposition within the can causes gas production and the can will swell or burst. However, there have been examples of poor manufacture (underprocessing) and poor hygiene allowing contamination of canned food by the obligate anaerobe *Clostridium botulinum*, which produces an acute toxin within the food, leading to severe illness or death. This organism produces no gas or obvious taste and remains undetected by taste or smell. Its toxin is denatured by cooking, though. Cooked mushrooms, handled poorly and then canned, can support the growth of Staphylococcus aureus, which produces a toxin that is not destroyed by canning or subsequent reheating.

Jellying

Food may be preserved by cooking in a material that solidifies to form a gel. Such materials include gelatine, agar, maize flour and arrowroot flour. Some foods naturally form a protein gel when cooked such as eels and elvers, and sipunculid worms which are a delicacy in the town of Xiamen in Fujian province of the People's Republic of China. Jellied eels are a delicacy in the East End of London where they are eaten with mashed potatoes. Potted meats in aspic, (a gel made from gelatine and clarified meat broth) were a common way of serving meat off-cuts in the UK until the 1950s. Many jugged meats are also jellied.

Potting

A traditional British way of preserving meat (particularly shrimp) is by setting it in a pot and sealing it with a layer of fat. Also common is potted chicken liver; compare pate.

Jugging

Meat can be preserved by jugging, the process of stewing the meat (commonly game or fish) in a covered earthenware jug or casserole. The animal to be jugged is usually cut into pieces, placed into a tightly-sealed jug with brine or gravy, and stewed. Red wine and/or the animal's own blood is sometimes added to the cooking liquid. Jugging was a popular method of preserving meat up until the middle of the 20th century.

Irradiation

Irradiation of food is the exposure of food to ionizing radiation; either high-energy electrons or X-rays from accelerators, or by gamma rays (emitted from radioactive sources as Cobalt-60 or Caesium-137). The treatment has a range of effects, including killing bacteria, molds and insect pests, reducing the ripening and spoiling of fruits, and at higher doses inducing sterility. The technology may be compared to pasteurization; it is sometimes called 'cold pasteurization', as the product is not heated. Irradiation is not effective against viruses or prions, it cannot eliminate toxins already formed by microorganisms, and is only useful for food of high initial quality.

The radiation process is unrelated to nuclear energy, but it may use the radiation emitted from radioactive nuclides produced in nuclear reactors. Ionizing radiation is hazardous to life (hence its usefulness in sterilisation); for this reason irradiation facilities have a heavily shielded irradiation room where the process takes place. Radiation safety procedures ensure that neither the workers in such facility nor the environment receive any radiation dose from the facility. Irradiated food does not become radioactive, and national and international expert bodies have declared food irradiation as wholesome. However, the wholesomeness of consuming such food is disputed by opponents and consumer organizations. National and international expert bodies have declared food irradiation as 'wholesome'; UN-organizations as WHO and FAO are endorsing to use food irradiation. International legislation on whether food may be irradiated or not varies worldwide from no regulation to full banning. Irradiation may allow lower quality or contaminated foodstuffs to be rendered marketable.

It is estimated that about 500,000 tons of food items are irradiated per year worldwide in over 40 countries. These are mainly spices and condiments with an increasing segment of fresh fruit irradiated for fruit fly quarantine.

Pulsed Electric Field Processing

Pulsed electric field (PEF) processing is a method for processing cells by means of brief pulses of a strong electric field. PEF holds potential as a type of low temperature alternative pasteurization process for sterilizing food products. In PEF processing, a substance is placed between two electrodes, then the pulsed electric field is applied. The electric field enlarges the pores of the cell membranes which kills the cells and releases their contents. PEF for food processing is a developing technology still being researched. There have been limited industrial applications of PEF processing for the pasteurization of fruit juices.

Modified Atmosphere

Modifying atmosphere is a way to preserve food by operating on the atmosphere around it. Salad crops which are notoriously difficult to preserve are now being packaged in sealed bags with an atmosphere modified to reduce the oxygen (O_2) concentration and increase the carbon dioxide (CO_2) concentration. There is concern that although salad vegetables retain their appearance and texture in such conditions, this method of preservation may not retain nutrients, especially vitamins. Grains may be preserved using carbon dioxide by one of two methods; either using a block of dry ice placed in the bottom and the can is filled with grain or the container can be purged from the bottom by gaseous carbon dioxide from a cylinder or bulk supply vessel. Carbon dioxide prevents insects, and depending on concentration, mold, and oxidation from damaging the grain. Grain stored in this way can remain edible for five years.

Nitrogen gas (N_2) at concentrations of 98% or higher is also used effectively to kill insects in grain through hypoxia . However, carbon dioxide has an advantage in this respect as it kills organisms through hypercarbia and depending on concentration hypoxia and, requiring concentrations of above 35% , or so. This makes carbon dioxide preferable for fumigation in situations where a hermetic seal cannot be maintained.

Burial in the Ground

Burial of food can preserve it due to a variety of factors: lack of light, lack of oxygen, cool temperatures, pH level, or desiccants in the soil. Burial may be combined with other methods such as salting or fermentation. Many root vegetables are very resistant to spoilage and require no other preservation than storage in cool dark conditions, for example by burial in the ground, such as in a storage clamp.

Century eggs are created by placing eggs in alkaline mud (or other alkaline substance) resulting in their "inorganic" fermentation through raised pH instead of spoiling. The fermentation preserves them and breaks down some of the complex, less flavourful proteins and fats into simpler more flavourful ones.

Most foods can be preserved in soil that is very dry and salty (thus a desiccant), or soil that is frozen.

Cabbage was traditionally buried in the fall in northern farms in the USA for preservation. Some methods keep it crispy while other methods produce sauerkraut. A similar process is used in the traditional production of kimchi. Sometimes meat is buried under conditions which cause preservation. If buried on hot coals or ashes, the heat can kill pathogens, the dry ash can desiccate, and the earth can block oxygen and further contamination. If buried where the earth is very cold, the earth acts like a refrigerator.

Controlled Use of Micro-organism

Some foods, such as many cheeses, wines, and beers will keep for a long time because their production uses specific micro-organisms that combat spoilage from other less benign organisms.

These micro-organisms keep pathogens in check by creating an environment toxic for themselves and other micro-organisms by producing acid or alcohol. Starter micro-organisms, salt, hops, controlled (usually cool) temperatures, controlled (usually low) levels of oxygen and/or other methods are used to create the specific controlled conditions that will support the desirable organisms that produce food fit for human consumption.

High Pressure Food Preservation

High pressure food preservation refers to high pressure used for food preservation. "Pressed inside a vessel exerting 70,000 pounds per square inch (480 MPa) or more, food can be processed so that it retains its fresh appearance, flavour, texture and nutrients while disabling harmful microorganisms and slowing spoilage." By 2001, adequate commercial equipment was developed so that by 2005 the process was being used for products ranging from orange juice to guacamole to deli meats and widely sold.

Food Fortification

Food fortification is the public health policy of adding micronutrients (essential trace elements and vitamins) to foodstuffs to ensure that minimum dietary requirements are met. Simple diets based on staple foods with little variation are often deficient in certain nutrients, either because they are not present in sufficient amounts in the soil of a region, or because of the inherent inadequacy of the diet. Addition of micronutrients to staples and condiments can prevent large-scale deficiency diseases in these cases.

Rationale

Several ranges of food supplements are recognised:

- additives which repair a deficit to "normal" levels
- additives which appear to enhance a food
- supplements taken in addition to the normal diet.

Many physicians today disagree with the premise that foodstuffs need supplementation, but accept that-for example-added calcium may provide benefit, or that adding folic acid may correct a nutritional deficiency especially in pregnant women. On a more controversial level, but well founded in scientific basis, is the science of using foods and food supplements to achieve a defined health goal. A common example of this use of food supplements is the extent to which body builders will use amino acid mixtures, vitamins and phytochemicals to enhance natural hormone production, increase muscle and reduce fat.

Moving on from this reasonably accepted usage, there is increasing evidence for the use of food supplements in established medical conditions. This nutritional supplementation using foods as medicine (nutraceuticals) has been effectively used in treating disorders affecting the immune system up to and including cancers. This goes beyond the definition of "food supplement", but should be included for the sake of completeness.

Food Supplements

There are several main groups of food supplements which can be considered:

- Vitamins and co-vitamins
- Essential minerals

- Essential fatty acids
- Essential amino acids
- Glyconutrients
- Phytonutrients
- Enzymes.

Examples of Fortified Foods

Iodised salt has been used in the United States since before World War II.

Folic acid is added to flour in many industrialized countries, and has prevented a significant number of neural tube defects in infants. It is, however, not uniform in its application, with more intake of folic acid through fortified flour among those who were already receiving high amounts through their diet.

Niacin has been added to bread in the USA since 1938 (when voluntary addition started), a programme which substantially reduced the incidence of pellagra.

Vitamin D is added to a few foods (especially margarine). Fluoride salts are added to water and toothpastes to prevent tooth decay. Water fluoridation is a controversial topic in some segments of the general public, although less so amongst established scientific bodies.

Calcium is frequently added to fruit juices, carbonated beverages and rice. "Golden rice" is a variety of rice which has been genetically modified to produce beta carotene.

Dietary Supplement

"Food supplement" redirects here. For food additions that alter the flavour, colour or longevity of food, see Food additive.

A dietary supplement, also known as food supplement or nutritional supplement, is a preparation intended to supplement the diet and provide nutrients, such as vitamins, minerals, fibre, fatty acids, or amino acids, that may be missing or may not be consumed in sufficient quantities in a person's diet. Some countries define dietary supplements as foods, while in others they are defined as drugs or natural health products.

Supplements containing vitamins or dietary minerals are included as a category of food in the Codex Alimentarius, a collection of internationally recognized standards, codes of practice, guidelines and other recommendations relating to foods, food production and food safety. These texts are drawn up by the Codex Alimentarius

Commission, an organization that is sponsored by the Food and Agriculture Organization of the United Nations (FAO) and the World Health Organization (WHO).

European Union

The European Union's Food Supplements Directive of 2002 requires that supplements be demonstrated to be safe, both in dosages and purity. Only those supplements that have been proven to be safe may be sold in the bloc without prescription. As a category of food, food supplements cannot be labeled with drug claims but can bear health claims and nutrition claims.

The dietary supplements industry in the United Kingdom (UK), one of the 27 countries in the bloc, strongly opposed the Directive. In addition, a large number of consumers throughout Europe, including over one million in the UK, and many doctors and scientists, had signed petitions by 2005 against what are viewed by the petitioners as unjustified restrictions of consumer choice. In 2004, along with two British trade associations, the Alliance for Natural Health had a legal challenge to the Food Supplements Directive referred to the European Court of Justice by the High Court in London. Although the European Court of Justice's Advocate General subsequently said that the bloc's plan to tighten rules on the sale of vitamins and food supplements should be scrapped, he was eventually overruled by the European Court, which decided that the measures in question were necessary and appropriate for the purpose of protecting public health. ANH, however, interpreted the ban as applying only to synthetically produced supplements—and not to vitamins and minerals normally found in or consumed as part of the diet. Nevertheless, the European judges did acknowledge the Advocate General's concerns, stating that there must be clear procedures to allow substances to be added to the permitted list based on scientific evidence. They also said that any refusal to add a product to the list must be open to challenge in the courts.

A survey of adults aged 18–64 years, conducted in 2001 in Ireland, the third-largest island in Europe, suggested that with the possible exception of niacin (flushing) and vitamin B6 (neuropathy), there appears to be little risk of the occurrence of adverse effects due to excessive consumption of vitamins in this population, based on current dietary practices.

Russia

The Russian Ministry of Health's order number 1898, dated 15 April 1997 and titled "Concerning the procedure for the examination and health certification of Active Dietary Supplements", states the following:

As a rule, BADSs are foodstuffs with clinically proven effectiveness. BADSs are recommended not only for prophylactics, but can be included into a complex therapy for the prevention of pharmaceutical therapy's side effects and for the achievement of complete remission.

The development of BADSs and their applications has been very fast moving. They were originally considered as dietary supplements for people who had heightened requirements for some normal dietary components (for example, sportsmen). Later, they were employed as preventive medicines against chronic diseases.

China

In China, prior to market entry, manufacturers must register dietary supplements with the State Food and Drug Administration (SFDA). The process requires investments of both time and money. After market entry, the registration is valid for five years and then must be renewed. The SFDA has a list of 27 approved functional claims that manufacturers of dietary supplements are allowed to make when registering a product. The process involves a strict testing protocol, including animal and human studies. To a large degree, it is similar to the approval of pharmaceuticals.

United States

In the United States, a dietary supplement is defined under the Dietary Supplement Health and Education Act of 1994 (DSHEA) as a product that is intended to supplement the diet and contains any of the following dietary ingredients:

- a vitamin
- a mineral
- a herb or other botanical (excluding tobacco)
- an amino acid
- a concentrate, metabolite, constituent, extract, or combination of any of the above.

Furthermore, it must also conform to the following criteria:

- intended for ingestion in pill, capsule, tablet, powder or liquid form
- not represented for use as a conventional food or as the sole item of a meal or diet
- labeled as a "dietary supplement".

The hormones DHEA (a steroid), pregnenolone (also a steroid) and the pineal hormone melatonin are marketed as dietary supplements in the US.

Regulation

The Food and Drug Administration (FDA) regulates dietary supplements as a category of foods, and not as drugs. While pharmaceutical companies are required to obtain FDA approval which involves assessing the risks and benefits prior to their entry into the market, dietary supplements do not need to be pre-approved by FDA before they can enter the market. Instead, manufacturers and distributors who wish to market dietary supplements that contain a "new dietary ingredient" (defined as "a vitamin; a mineral; a herb or other botanical; an amino acid; a dietary substance for use by man to supplement the diet by increasing total dietary intake; or a concentrate, metabolite, constituent, extract, or combination of any of the above dietary ingredients" not marketed before October 15, 1994) must notify the FDA beforehand.

The notification requires information indicating the ingredient is safe, and the ingredient can not be marketed (sold or delivered for sale) for seventy-five days following filing the information. During this time the agency reviews the information for adequacy and safety concerns; fifteen days after the this period (ninety days after the information was filed) the FDA posts nonproprietary information on their website. Listing the information means the ingredient can be marketed, but does not mean it is necessarily safe. On September 24, 2007 the FDA has implemented a "current good manufacturing practices" policy to ensure dietary supplements "are produced in a quality manner, do not contain contaminants or impurities, and are accurately labeled" and covers the manufacturing, packaging, labelling and storing of supplements, with requirements for quality control, design and construction of manufacturing plants, testing of ingredients and final products, record keeping and complaints processes.

The DSHEA, passed in 1994, was the subject of lobbying efforts by the manufacturers of dietary supplements and restricted the ability of the FDA to exert authority over supplements so long as manufacturers made no claims about their products treating, preventing or curing diseases.

According to *Consumer Reports*, "The law has left consumers without the protections surrounding the manufacture and marketing of over-the-counter or prescription medications" and it became the FDA's responsibility to prove that a supplement wasn't safe. While pharmaceutical manufacturers must demonstrate their products are effective as well as being safe, supplement manufacturers are not required to demonstrate efficacy.

The FDA has only ever found one dietary supplement to be unsafe, the weight loss/energy supplement ephedra. Discussing the legislation, *Time* referred to the DSHEA as "ill-conceived and reprehensible", that "gives the industry virtually free reign [*sic*] to market products defined as dietary supplements, while severely limiting the FDA's ability to regulate them". The DSHEA was heavily lobbied for by the supplement industry, and was criticized for exposing the public to worthless compounds that bilked consumers out of money to no benefit. Because of the requirements put into place by the DSHEA, the FDA must demonstrate that individual supplements are unsafe using their adverse events reporting system, which it is estimated captures only 1% of all adverse events linked to supplements. The FDA has also lacked the funding to undertake the rigorous tests to meet the standards for a supplement to be considered "hazardous" and thus removed from the market; in the one situation where this standard was reached (ephedra), the agency faced significant opposition from the supplement industry and the United States Congress, instead limiting themselves to making announcements about problematic supplement safety records on their website.

A 2001 study, published in *Archives of Internal Medicine*, found broad public support for greater governmental regulation of dietary supplements than was currently permitted by DSHEA. The researchers found that a majority of Americans supported pre-marketing approval by the FDA, increased oversight of harmful supplements, and greater scrutiny of the truthfulness of supplement label claims.

Quality

Under the FDA's final rule on good manufacturing practices, quality is defined as meaning "that the dietary supplement consistently meets the established specifications for identity, purity, strength, and composition and has been manufactured, packaged, labeled, and held under conditions to prevent adulteration under section 402 (a)(1), (a)(2), (a)(3), and (a)(4) of the Federal Food, Drug, and Cosmetic Act". The new regulations allow FDA inspectors to look at a company's records upon request. However, enforcement could be difficult given the number of supplement manufacturers and the 16% decline in FDA investigators from 2003 to 2006. Much of the contamination is due to poor raw ingredients. Suppliers provide certificates of analysis stating that they have tested the material. Under the 2003 proposed rule, manufacturers would have been required to retest the supplied ingredients. Under the final rule, testing for identity is always required. Other retesting is not required if the manufacturer has verified the reliability of the ingredient supplier.

In the U.S., contamination and false labelling are "not uncommon". Independent certification programs exist, but these may have problems as well. United States Pharmacopeia manages the Dietary Supplement Verification Program (DSVP). Its USP Verified Mark seal indicates that the product has been tested for integrity, purity, dissolution, and safe manufacturing, and it is the only certification program which conducts random off-the-shelf testing.

The USP program will not certify products which contain ingredients that the USP's Dietary Supplement Information Expert Committee determines have a safety risk. ConsumerLab.com randomly tests some dietary supplements and makes the results available to subscribers. It has reported that 25% of the supplements it tests have problems, and for multivitamins about half had problems. In 2008 ConsumerLab criticized the USP for proposing a 10 microgram perdaily serving limit on lead in dietary supplements and drugs. It noted that under the FDA's 2006 guidance on lead in candy, only 0.2 micrograms of lead per serving are allowed. NSF International, HFL Sport Science, and the Natural Products Association also have a dietary supplement certification programs.

Permissible Claims

If a dietary supplement claims to cure, mitigate, or treat a disease, it would be considered an unauthorized new drug and in violation of the applicable regulations and statutes. As the FDA states, in a response to this question, in a FAQ:

Is it legal to market a dietary supplement product as a treatment or cure for a specific disease or condition?

No, a product sold as a dietary supplement and promoted on its label or in labelling as a treatment, prevention or cure for a specific disease or condition would be considered an unapproved—and thus illegal—drug. To maintain the product's status as a dietary supplement, the label and labelling must be consistent with the provisions in the Dietary Supplement Health and Education Act (DSHEA) of 1994.*

*Labelling refers to the label as well as accompanying material that is used by a manufacturer to promote and market a specific product.

Dietary supplements are permitted to make structure/function claims. These are broad claims that the product can support the structure or function of the body (e.g., "glucosamine helps support healthy joints", "the hormone melatonin helps establish normal sleep patterns"). The FDA must be notified of these claims within 30 days of their first use, and there is a requirement that these claims be substantiated. In reality, misleading claims about supplements are common, particularly on poorly regulated commercial websites. For example, the compound hydrazine sulfate is sold as a dietary supplement in the USA and promoted as a treatment for cancer, despite little evidence that it is either safe or effective.

Other claims that required approval from FDA include health claims and qualified health claims. Health claims are permitted to be made if they meet the requirements for the claims found in the applicable regulations. Qualified health claims can be made through a petition process, including scientific information, if FDA has not approved a prior petition.

Chapter 2

Application of Modern Biotechnology in Food Processing

Contrary to its name, biotechnology is not a single technology. Rather it is a group of technologies that share two (common) characteristics — working with living cells and their molecules and having a wide range of practice uses that can improve our lives.

Biotechnology can be broadly defined as "using organisms or their products for commercial purposes." As such, (traditional) biotechnology has been practices since he beginning of records history. (It has been used to:) bake bread, brew alcoholic beverages, and breed food crops or domestic animals. But recent developments in molecular biology have given biotechnology new meaning, new prominence, and new potential. It is (modern) biotechnology that has captured the attention of the public. Modern biotechnology can have a dramatic effect on the world economy and society.

One example of modern biotechnology is genetic engineering. Genetic engineering is the process of transferring individual genes between organisms or modifying the genes in an organism to remove or add a desired trait or characteristic. Examples of genetic engineering are described later in this document. Through genetic engineering, genetically modified crops or organisms are formed. These GM crops or GMOs are used to produce biotech-derived foods. It is this specific type of modern biotechnology, genetic engineering, that seems to generate the most attention and concern by consumers and consumer groups. What is interesting is that modern biotechnology is far more precise than traditional forms of biotechnology and so is viewed by some as being far safer.)

How does Modern Biotechnology Work?

All organisms are made up of cells that are programmed by the same basic genetic material, called DNA (deoxyribonucleic acid). Each unit of DNA is made up of a combination of the following nucleotides — adenine (A), guanine (G), thymine (T), and cytosine (D) — as well as a sugar and a phosphate. These nucleotides pair up into strands that twist together into a spiral structure call a "double helix." This double helix is DNA. Segments of the DNA tell individual cells how to produce specific proteins. These segments are genes. It is the presence or absence of the specific protein that gives an organism a trait or characteristic. More than 10,000 different genes are found in most plant and animal species. This total set of genes for an organism is organized into chromosomes within the cell nucleus. The process by which a multicellular organism develops from a single cell through an embryo stage into an adult is ultimately controlled by the genetic information of the cell, as well as interaction of genes and gene products with environmental factors.

When cells reproduce, the DNA strands of the double helix separate. Because nucleotide A always pairs with T and G always pairs with C, each DNA strand serves as a precise blueprint for a specific protein. Except for mutations or mistakes in the replication process, a single cell is equipped with the information to replicate into millions of identical cells. Because all organisms are made up of the same type of genetic material (nucleotides A, T, G, and C), biotechnologists use enzymes to cut and remove DNA segments from one organism and recombine it with DNA in another organism. This is called recombinant DNA (rDNA) technology, and it is one of the basic tools of modern biotechnology. rDNA technology is the laboratory manipulation of DNA in which DNA, or fragments of DNA from different sources, are cut and recombined using enzymes. This recombinant DNA is then inserted into a living organism. rDNA technology is usually used synonymously with genetic engineering. rDNA technology allows researchers to move genetic information between unrelated organisms to produce desired products or characteristics or to eliminate undesirable characteristics.

Genetic engineering is the technique of removing, modifying or adding genes to a DNA molecule in order to change the information it contains. By changing this information, genetic engineering changes the type or amount of proteins an organism is capable of producing. Genetic engineering is used in the production of drugs, human gene therapy, and the development of improved plants. For example, an "insect protection" gene (Bt) has been inserted into several crops - corn, cotton, and potatoes - to give farmers new tools for integrated pest management. Bt corn is resistant to European corn borer. This inherent resistance thus reduces a farmers pesticide use for controlling European corn borer, and in turn requires less chemicals and potentially provides higher yielding Agricultural Biotechnology.

Although major genetic improvements have been made in crops, progress in conventional breeding programs has been slow. In fact, most crops grown in the US produce less than their full genetic potential. These shortfalls in yield are due to the inability of crops to tolerate or adapt to environmental stresses, pests, and diseases. For example, some of the world's highest yields of potatoes are in Idaho under irrigation, but in 1993 both quality and yield were severely reduced because of cold, wet weather and widespread frost damage during June. Some of the world's best bread wheats and malting barleys are produced in the north-central states, but in 1993 the disease Fusarium caused an estimated $1 billion in damage. Scientists have the ability to insert genes that give biological defense against diseases and insects, thus reducing the need for chemical pesticides, and they will soon be able to convey genetic traits that enable crops to better withstand harsh conditions, such as drought. The International Laboratory for Tropical Agricultural Biotechnology (ILTAB) is developing transformation techniques and applications for control of diseases caused by plant viruses in tropical plants such as rice, cassava and tomato. In 1995, ILTAB reported the first transfer through biotechnology of a resistance gene from a wild species of rice to a susceptible cultivated rice variety. The transferred gene expressed resistance to Xanthomonas oryzae, a bacterium which can destroy the crop through disease. The resistant gene was transferred into susceptible rice varieties that are cultivated on more than 24 million hectares around the world.

Benefits can also be seen in the environment, where insect-protected biotech crops reduce the need for chemical pesticide use. Insect-protected crops allow for less potential exposure of farmers and ground-water to chemical residues, while providing farmers with season-long control. Also by reducing the need for pest control, impacts and resources spent on the land are less, thereby preserving the topsoil.

Major advances also have been made through conventional breeding and selection of livestock, but significant gains can still be made by using biotechnology. Currently, farmers in the U.S spend $17 billion dollars on animal health. Diseases such as hog cholera and pests such as screwworm have been eradicated. Uses of biotechnology in animal production include development of vaccines to protect animals from disease, production of several calves from one embryo (cloning), increase of animal growth rate, and rapid disease detection.

Modern biotechnology has offered opportunities to produce more nutritious and better tasting foods, higher crop yields and plants that are naturally protected from disease and insects. Modern biotechnology allows for the transfer of only one or a few desirable genes, thereby permitting scientists to develop crops with specific beneficial traits and reduce undesirable traits.

Traditional biotechnology such as cross-pollination in corn produces numerous, non-selective changes. Genetic modifications have produced fruits that can ripen on the vine for better taste, yet have longer shelf lives through delayed pectin degradation. Tomatoes and other produce containing increased levels of certain nutrients, such as vitamin C, vitamin E, and or beta carotene, and help protect against the risk of chronic diseases, such as some cancers and heart disease. Similarly introducing genes that increase available iron levels in rice three-fold is a potential remedy for iron deficiency, a condition that effects more than two billion people and causes anemia in about half that number. Most of the today's hard cheese products are made with a biotech enzyme called chymosin. This is produced by genetically engineered bacteria which is considered more purer and plentiful than it's naturally occurring counterpart, rennet, which is derived from calf stomach tissue.

In 1992, Monsanto Company successfully inserted a gene from a bacterium into the Russet Burbank potato. This gene increases the starch content of the potato. Higher starch content reduces oil absorption during frying, thereby lowering the cost of processing french fries and chips and reducing the fat content in the finished product. This product is still awaiting final development and approval.

Modern biotechnology offers effective techniques to address food safety concerns. Biotechnical methods may be used to decrease the time necessary to detect foodborne pathogens, toxins, and chemical contaminants, as well as to increase detection sensitivity. Enzymes, antibodies, and microorganisms produced using rDNA techniques are being used to monitor food production and processing systems for quality control.

Biotechnology can compress the time frame required to translate fundamental discoveries into applications. This is done by controlling which genes are altered in an organized fashion. For example, a known gene sequence from a corn plant can be altered to improve yield, increase drought tolerance, and produce insect resistance (Bt) in one generation. Conventional breeding techniques would take several years. Conventional breeding techniques would require that a field of corn is grown and each trait is selected from individual stalks of corn. The ears of corn from selected stalks with each desired trait (e.g, drought tolerance and yield performance) would then be grown and combined (cross-pollinated). Their offspring (hybrid) would be further selected for the desired result (a high performing corn with drought tolerance). With improved technology and knowledge about agricultural organisms, processes, and ecosystems, opportunities will emerge to produce new and improved agricultural products in an environmentally sound manner.

In summary, modern biotechnology offers opportunities to improve product quality, nutritional content, and economic benefits. The genetic make-up of plants and animals can be modified by either insertion of new useful genes or removal of unwanted ones. Biotechnology is changing the way plants and animals are grown, boosting their value to growers, processors, and consumers.

Industrial Biotechnology

Industrial biotechnology applies the techniques of modern molecular biology to improve the efficiency and reduce the environmental impacts of industrial processes like textile, paper and pulp, and chemical manufacturing. For example, industrial biotechnology companies develop biocatalysts, such as enzymes, to synthesize chemicals. Enzymes are proteins produced by all organisms. Using biotechnology, the desired enzyme can be manufactured in commercial quantities.

Commodity chemicals (e.g., polymer-grade acrylamide) and speciality chemicals can be produced using biotech applications. Traditional chemical synthesis involves large amounts of energy and often-undesirable products, such as HCl. Using biocatalysts, the same chemicals can be produced more economically and more environmentally friendly. An example would be the substitution of protease in detergents for other cleaning compounds. Detergent proteases, which remove protein impurities, are essential components of modern detergents.

They are used to break down protein, starch, and fatty acids present on items being washed. Protease production results in a biomass that in turn yields a useful by-product- an organic fertilizer. Biotechnology is also used in the textile industry for the finishing of fabrics and garments. Biotechnology also produces biotech-derived cotton that is warmer, stronger, has improved dye uptake and retention, enhanced absorbency, and wrinkle- and shrink-resistance. Some agricultural crops, such as corn, can be used in place of petroleum to produce chemicals. The crop's sugar can be fermented to acid, which can be then used as an intermediate to produce other chemical feedstocks for various products. It has been projected that 30% of the world's chemical and fuel needs could be supplied by such renewable resources in the first half of the next century. It has been demonstrated, at test scale, that biopulping reduces the electrical energy required for wood pulping process by 30%.

Environmental Biotechnology

Environmental biotechnology is the used in waste treatment and pollution prevention. Environmental biotechnology can more efficiently clean up many wastes than conventional methods and greatly reduce our dependence on methods for land-based disposal.

Every organism ingests nutrients to live and produces by-products as a result. Different organisms need different types of nutrients. Some bacteria thrive on the chemical components of waste products. Environmental engineers use bioremediation, the broadest application of environmental biotechnology, in two basic ways. They introduce nutrients to stimulate the activity of bacteria already present in the soil at a waste site, or add new bacteria to the soil. The bacteria digest the waste at the site and turn it into harmless by-products. After the bacteria consume the waste materials, they die off or return to their normal population levels in the environment.

Bioremediation, is an area of increasing interest. Through application of biotechnical methods, enzyme bioreactors are being developed that will pretreat some industrial waste and food waste components and allow their removal through the sewage system rather than through solid waste disposal mechanisms. Waste can also be converted to biofuel to run generators. Microbes can be induced to produce enzymes needed to convert plant and vegetable materials into building blocks for biodegradable plastics.

In some cases, the by-products of the pollution-fighting microorganisms are themselves useful. For example, methane can be derived from a form of bacteria that degrades sulphur liquor, a waste product of paper manufacturing. This methane can then be used as a fuel or in other industrial processes.

Human Applications

Biotechnical methods are now used to produce many proteins for pharmaceutical and other specialized purposes. A harmless strain of

Escherichia coli bacteria, given a copy of the gene for human insulin, can make insulin. As these genetically modified (GM) bacterial cells age, they produce human insulin, which can be purified and used to treat diabetes in humans. Microorganisms can also be modified to produce digestive enzymes. In the future, these microorganisms could be colonized in the intestinal tract of persons with digestive enzyme insufficiencies. Products of modern biotechnology include artificial blood vessels from collagen tubes coated with a layer of the anticoagulant heparin.

Gene therapy – altering DNA within cells in an organism to treat or cure a disease – is one of the most promising areas of biotechnology research. New genetic therapies are being developed to treat diseases such as cystic fibrosis, AIDS and cancer. DNA fingerprinting is the process of cross matching two strands of DNA. In criminal investigations, DNA from samples of hair, bodily fluids or skin at a crime scene are compared with those obtained from the suspects. In practice, it has become one of the most powerful and widely known applications of biotechnology today. Another process, polymerase chain reaction (PCR), is also being used to more quickly and accurately identify the presence of infections such as AIDS, Lyme disease and Chlamydia.

Paternity determination is possible because a child's DNA pattern is inherited, half from the mother and half from the father. To establish paternity, DNA fingerprints of the mother, child and the alleged father are compared. The matching sequences of the mother and the child are eliminated from the child's DNA fingerprint; what remains comes from the biological father. These segments are then compared for a match with the DNA fingerprint of the alleged father.

DNA testing is also used on human fossils to determine how closely related fossil samples are from different geographic locations and geologic areas. The results shed light on the history of human evolution and the manner in which human ancestors settled different parts of the world.

Biotechnology for the 21st Century

Experts in United States anticipate the world's population in 2050 to be approximately 8.7 billion persons. The world's population is growing, but its surface area is not. Compounding the effects of population growth is the fact that most of the earth's ideal farming land is already being utilized. To avoid damaging environmentally sensitive areas, such as rain forests, we need to increase crop yields for land currently in use. By increasing crop yields, through the use of biotechnology the constant need to clear more land for growing food is reduced.

Countries in Asia, Africa, and elsewhere are grappling with how to continue feeding a growing population. They are also trying to benefit more from their existing resources. Biotechnology holds the key to increasing the yield of staple crops by allowing farmers to reap bigger harvests from currently cultivated land, while preserving the land's ability to support continued farming.

Malnutrition in underdeveloped countries is also being combated with biotechnology. The Rockefeller Foundation is sponsoring research on "golden rice", a crop designed to improve nutrition in the developing world. Rice breeders are using biotechnology to build Vitamin A into the rice. Vitamin A deficiency is a common problem in poor countries. A second phase of the project will increase the iron content in rice to combat anemia, which is widespread problem among women and children in underdeveloped countries. Golden rice, expected to be for sale in Asia in less than five years, will offer dramatic improvements in nutrition and health for millions of people, with little additional costs to consumers.

Similar initiatives using genetic manipulation are aimed at making crops more productive by reducing their dependence on pesticides, fertilizers and irrigation, or by increasing their resistance to plant diseases. Increased crop yield, greater flexibility in growing environments, less use of chemical pesticides and improved nutritional content make agricultural biotechnology, quite literally, the future of the world's food supply.

Concerns about Biotechnology

As biotechnology has become widely used, questions and concerns have also been raised. The most vocal opposition has come from European countries. One of the main areas of concern is the safety of genetically engineered food.

In assessing the benefits and risks involved in the use of modern biotechnology, there are a series of issues to be addressed so that informed decisions can be made. In making value judgments about risks and benefits in the use of biotechnology, it is important to distinguish between technology-inherent risks and technology-transcending risks. The former includes assessing any risks associated with food safety and the behaviour of a biotechnology-based product in the environment. The latter involve the political and social context in which the technology is used, including how these uses may benefit or harm the interests of different groups in society.

The health effects of foods grown from genetically engineered crop depend on the composition of the food itself. Any new product may have either beneficial or occasional harmful effects on human health.

For example, a biotech-derived food with a higher content of digestible iron is likely to have a positive effect if consumed by iron-deficient individuals. Alternatively, the transfer of genes from one species to another may also transfer the risk for exposure to allergens.

These risks are systematically evaluated by FDA and identified prior to commercialization. Individuals allergic to certain nuts, for example, need to know if genes conveying this trait are transferred to other foods such as soybeans. Labelling would be required if such crops were available to consumers.

Among the potential ecological risks identified are increased weediness, due to cross- pollination from genetically modified crops spreads to other plants in nearby fields. This may allow the spread of traits such as herbicide-resistance to non-target plants that could potentially develop into weeds. This ecological risk is assessed when deciding if a plant with a given trait should be released into a particular environment, and if so, under what conditions.

Other potential ecological risks stem from the use of genetically modified corn and cotton with insecticidal genes from Bacillus thuringiensis (Bt genes). This may lead to the development of resistance to Bt in insect populations exposed to the biotech-derived crop. There also may be risks to non-target species, such as birds and butterflies, from the plants with Bt genes. The monitoring of these effects of new crops in the environment and implementation of effective risk management approaches is an essential component of further research. It is also important to keep all risks in perspective by comparing the products of biotechnology and conventional agriculture.

The reduction of biodiversity would represent a technology-transcending risk. Reduced biological diversity due to destruction of tropical forests, conversion of land to agriculture, overfishing, and the other practices to feed a growing world population is a significant loss far more than any potential loss of biodiversity due to biotech-derived crop varieties. Improved governance and international support are necessary to limit loss of biodiversity.

What we know from our understanding of science and more than a decade of experience with biotech-derived plants is the following: There is no evidence that genetic transfers between unrelated organisms pose human health concerns that are different from those encountered with any new plant or animal variety. The risks associated with biotechnology are the same as those associated with plants and microbes developed by conventional methods.

Consumer and Food Industry Perspectives

Survey research over the past decade shows that biotechnology is not likely to become an important issue for most American consumers. Consumers find biotechnology acceptable when they believe it offers benefits and it is safe. Surveys have consistently found that a majority of American consumers are willing to buy insect-protected food crops developed through biotechnology that use fewer chemical pesticides, as well as more nutritious foods. American consumers also appreciate the role that biotechnology can play in feeding the world. Research shows that European consumers are much less supportive of all biotechnology applications.

Surveys since 1992 show that relatively few U.S. consumers have heard or read much about biotechnology. News about the cloned sheep pushed awareness to 50 percent in March 1997. Surveys in the first three months of 2000 show that awareness has fallen back to just over one-third in the United States. Such trends reflect the fact that most people get their information about biotechnology from the media. Unfortunately, many consumers also do not understand some fundamental principles of biology. European consumer awareness is somewhat higher, but knowledge is still low.

Media coverage in the United States has generally been balanced (which helps account for our relatively high levels of acceptance). This is in sharp contrast to the European media, which have played upon fear of the unknown. The European media have also tended to accept opponents' claims without question. Another issue is that many people no longer have a connection to agriculture. In fact, research has shown that many consumers are unaware that all foods are derived from plants or animals that already have been genetically modified through traditional (but imprecise) breeding methods. American consumers look to health professionals and scientific experts for credible information, but place relatively little trust in the activists who oppose biotechnology. Research shows that acceptance increases significantly when American consumers learn that organizations such as the National Academy of Sciences and the U.S. Food and Drug Administration have determined that biotech-derived foods are safe. In contrast, European consumers express the most trust in those groups that oppose biotechnology. They have much less confidence in government, industry, or even scientists. American culture is more supportive and rewarding of new technology.

Europeans tend to view food differently from U.S. consumers. In fact, some Europeans reject all American food products. Europeans also want to protect their small farms to maintain open space and rural employment. Such forces underlie much of the European anxiety about agricultural biotech - especially since it is seen as an "American invention."

Most of the industry leaders interviewed are quite enthusiastic about the benefits of biotechnology — especially in terms of increased food availability, enhanced nutrition, and environmental protection. Most feel that biotechnology has already provided benefits to consumers. Almost all recognize that foods developed through biotechnology have already been part of consumers' everyday diet. They clearly do not agree with most of the opponents' claims and tend to have almost no trust in such groups. Their main concerns involve lack of consumer acceptance — not the safety of the foods. They express high levels of confidence in the science and the regulatory process. In fact, almost none feel that biotechnology should not be used because of uncertain, potential risks. Most food industry leaders do not feel it is necessary to have special labels on biotech-derived foods. They express concerns that such labels would be perceived as a warning by consumers. They also worried that the need to segregate commodities would pose financial and logistical burdens on everyone in the system - including consumers. Food industry leaders recognize a major need to educate the public about biotechnology. They look to third parties, such as university and government scientists to provide such leadership.

Research shows that consumers will accept biotech foods if they see a benefit to themselves or society and if the price is right. Their responses to foods developed through biotechnology are basically the same as for any other food - taste, nutrition, price, safety and convenience are the major factors that influence our decisions about which foods to eat. How seeds and food ingredients are developed will only be relevant for a relatively small group of concerned, consumers. The potential for public concerns has led several food companies to change their products to avoid biotech-derived ingredients. For example, Gerber Foods received threats from Green peace because they had determined the company was using biotech-derived food ingredients (mainly soy). The company firmly believes that the biotech foods are safe to consume. Gerber agreed to drop some of its existing corn and soybean suppliers in favor of ones that can produce crops that are not genetically altered. It became an issue that is suddenly confronting other food companies.

A private manufacturer in California, called Healthy Times Natural Food has switched from Canola oil (which sometimes is genetically modified) to safflower oil after facing questions from Green peace. The controversy is due in part to the fact that the organic industry is using public concern as a tool for marketing their products as free of biotech ingredients.

National and International Biotechnology Policy

National governments and international policy making bodies rely on food scientists and others to develop innovations that will create marketable food products and increase food supplies.

Governments also rely on scientific research because they are responsible for setting health and safety standards regarding new developments. International organizations can suggest policy approaches and help develop international treaties that are ratified by national governments. Economic success in the competitive international market demands that food production become more efficient and profitable. National governments and international organizations support food biotechnology as a means to avoid global food shortages. Many policy making bodies are also trying to balance support of the food biotechnology industry with public calls for their regulation. Such regulations are necessary to protect public health and safety, to promote international trade, conserve natural resources, and account for ethical issues.

The majority of processed foods on the market contain soy or corn ingredients that come from GM plants. To date none have posed a food safety risk. The chief safety concerns are the potential to alter nutrient content or introduce allergens. Federal agencies involved in biotechnology regulation include the U.S. Department of Agriculture (USDA) which evaluates agricultural production processes for all foods; the Food and Drug Administration (FDA), which evaluates whole non-animal foods (seafood), food ingredients, and food additives; and the Environmental Protection Agency (EPA), which evaluates plants with insecticidal properties.

Developers of GM plants and biotech-derived foods are required to consult with FDA prior to the commercialization of the product. This consultation procedure entails a science-based safety assessment of the product that focuses on the protection of the consumer, developer, and the environment. Thus developers, have a strong incentive to cooperate fully with FDA and the other agencies prior to marketing their products.

Labelling Issues

Labelling food derived from GM plants and animals is an important, but complex issue. Some consumers and consumer groups believe they have a right to know whether biotechnology was used to produce food. Others believe labelling is not necessary if foods are essentially equivalent in composition. Food labels are regulated by FDA and, in some cases, by USDA. Regulatory agencies are concerned with ensuring that food labels are both true and not misleading. In accordance with their statutory mandate, the FDA has determined that a food product should be labelled as a product of biotechnology only if it has been changed in some significant way. The U.S. Court of Appeals recently reaffirmed that the FDA's labelling policy is correct. Based on the input from three public hearings, the FDA has recognized that the most effective way to allow for informed choice will be a system of voluntary labelling for foods not produced through biotechnology. If the demand for "GM-free" food is real, a market will become viable.

In this case, meaningful choice can be provided to concerned consumers without imposing costs on or denying benefits to the majority of consumers, who support biotechnology. This may be the only viable and cost-effective option for the food industry (given the costs and difficulties associated with segregation and identity preservation.)

National surveys conducted with American consumers between 1997 and 2000 found about three-quarters of consumers supported this FDA labelling policy. There is evidence from focus groups that US consumers are already overwhelmed by the level of detail on food labels and mainly look for relevant information about nutrition. Recent focus groups have demonstrated that the wording on labels has a significant effect on consumer understanding and acceptance of biotechnology. Any label messages need to be simple and clear. However, labelling will be counterproductive without education.

It is important to understand when FDA will and won't require a product label. The same labelling laws that apply to all other foods and food ingredients apply to products of food biotechnology. Requirements for all food labels mandate proper identification of products and notice of health or safety concerns. Labelling would be required in the following instances – but not simply because the products were made using biotechnology.

- Potential food allergy is an example of a health or safety risk that would mandate a product label (e.g., if a gene has been transferred from a known allergen, such as a peanut).
- When substantial changes have been made to the expected nutritional composition of a food (e.g., vitamins have been added to a particular type of produce).
- If the common identity of a food product has been changed (e.g., if the composition of a fruit or vegetable becomes significantly different it would have to have a different name).

Labelling of biotech-derived foods will be required in some countries (such as Japan and the European Union) for cultural and religious reasons. These countries also believe labelling should be required simply because the consumers want to know more about the content of the food or how it was produced. Some have suggested that such labelling initiatives have been driven by economic and political interests rather than by scientific need. Consumers in these countries will ultimately end up paying even higher prices for their food as a result of such strict labelling policies.

The applications of biotechnology are so broad, and the advantages so compelling, that virtually every industry is using this technology. Developments are underway in areas as diverse as pharmaceuticals, diagnostics, textiles, aquaculture, forestry, chemicals, household products, environmental cleanup, food processing and forensics to name a few. Biotechnology is enabling these industries to make new or better products, often with greater speed, efficiency and flexibility. Biotechnology holds significant promise to the future but certain amount of risk is associated with any area. Biotechnology must continue to be carefully regulated so that the maximum benefits are received with the least risk. Biotechnology is at a crossroads in terms of public acceptance. Many Americans have not yet formed a solid opinion on this complex issue. International developments over the next few years will certainly have a major influence on the long-term viability of biotechnology. The future of the world food supply depends upon how well scientists, government, and the food industry are able to communicate with consumers about the benefits and safety of the technology.

Several major initiatives are under way to strengthen the regulatory process and to communicate more effectively with consumers. Both the USDA and FDA have opened their regulatory systems to outside review and public comment. The biotechnology industry, university scientists and others are also conducting educational programs.

These should further strengthen consumer confidence. This partnership among the public and private sectors will support these emerging technologies that will prove vital to the U.S. economy and the developing world in the new millennium. Even Europe will soon find the real benefits of biotechnology compelling.

The 6 Top Trends in Food Processing

When it comes to food and beverage products, one rule of thumb defines a true trend: Real trends don't come and go; they grow – over years, decades or even longer. For processors, fads can still bring in big money, but the risk is high. Ask any food executive caught last year with a million bucks' worth of low-carb inventory.

In between fads and trends are trendlets. These are the bubbles that pop up within a trend and are worth noting because they can provide a hook for food and beverage processors to hang their R&D hats on.

We went shopping for real trends. We sifted through numerous media reports on what was hot, comments from the editorial advisory boards of both *Food Processing* and our sister magazine *Wellness Foods* and the wisdom shared at the many food shows we attend each year. We culled a huge list down to six essential trends.

New Food Products Resource Center

Food Processing reports on a variety of new food products. Find consumer market research for particular products, monthly product roll-out roundups and other monthly features in one place.

We've pegged Organic as the biggest and most significant trend for processors to watch. No great surprise, but more important than our No. 2 pick, Health & Wellness? Yes. Now entering its second generation of double-digit growth, organic is on another big upswing.

In simple terms, although the Health & Wellness category has a larger footprint than Organic – and the two often are lumped together erroneously – the fact is, any food product can be formulated to sport an organic label. However, not all products can wear the "healthy" tag.

Example: A rapidly growing niche in the beverage world is that of organic beer, wine and spirits. Discounting the studies showing possible benefits of moderate alcohol intake, organic vodka won't likely get a health claim. Health & Wellness is the second and broadest of the hot trends. It encompasses such huge components as diabetes and obesity, kids' health, food safety, women's health, allergies and immunity as well as the fringe issues of "well-being" and "energy."

As a trend, Age Awareness certainly overlaps with Health & Wellness, especially as the latter concerns our aging population. But there are numerous non-health aspects for processors to consider as they help our 77 million baby boomers segue into their dotage.

Portion Control is our No. 4 trend. In some ways, it's just a health tool. But this year, especially, it deserves its own category because it constitutes a merging of health with the perennial trend for more convenience. And it's one trend numerous processors from all categories are jumping on.

Globalization is No. 5. Immigration controversies aside, Asian, Hispanic, African American and other ethnic minorities will make up more than 35 percent of the U.S. population in about five years, according to estimates derived from U.S. Census Bureau data. About half that figure will be Hispanics. But globalization and ethnic influences are more than population figures. Today's businesses are international more often than not. The cultural traffic and instant global information (via mass media and the Internet) mean rapid diffusion of once regional preferences.

Rounding out the six major processing trends are Kosher and halal certification. The strange bedfellows continue to be prolific growth areas, with Kosher still progressing at double-digit rates and halal experiencing a sudden and major growth spurt.

Organic = Healthy

That equation is not necessarily true, but the message is so ingrained in the minds of millions of consumers that the math cannot be ignored.

U.S. organic food sales totaled nearly $14 billion in 2005, according to the Organic Trade Assn, Greenfield, Mass. (OTA). Although this represents a mere 2.5 percent of all retail food sales, that total is a 31 percent increase over 2003 figures. According to OTA, sales of organic foods are expected to reach nearly $16 billion by the end of 2006.

"These findings show there is continued strong growth for organic products," says Caren Wilcox, OTA's executive director. OTA statistics show organic food categories experiencing the greatest growth during 2005 included meat (55.4 percent), condiments (24.2 percent) and dairy products (23.5 percent). Even Fido and Garfield are going organic: One of the fastest-growing organic categories during 2005 was pet food (46 percent).

"We're at a point where demand for organic product exceeds supply," says Robert Schueller, director of public relations for Melissa's/World Variety Produce Inc., Los Angeles. "Organic everything is hot, and it's not just produce. Most of the top retailers in the country offer organic products in their stores, and many foodservice establishments have taken notice, offering organic foods and ingredients more often on the menu."

One of the largest distributors of variety organic produce in the country, Melissa's offers more than 350 organic produce items. Gearing up for a larger-than-normal increase in demand, Schueller notes the number will reach about 400 SKUs by year's end. Since the 2004 start of its Good Life Food brand of organic processed items, sales have grown an average of 20 percent a year. Forecasts double that figure for 2006.

It's not just the patchouli set that sees the advantages to organic foods. Surveys show the majority of Americans are concerned about what's in their foods, where those foods come from and potential health risks from pesticides and chemicals in the food chain.

David Johnson, president-North America commercial at Kraft Foods Inc., Northfield, Ill, has been quoted describing the organic food trend as "a freight train that's going to pick up steam." As the second largest food company in North America, the company is in the position to help make this a self-fulfilling prophecy.

According to a recent study by ACNielsen, Chicago, organic products topped the list of "best performing" items in the "good-for-you" product segments. But organic is its own trend, extending beyond foods, beverages and pet foods. The category has acquired such trendlets as environmental consciousness and sustainability, Fair Trade, local production, energy conservation and "natural," minimally processed or stripped-down formulations. Although not making an organic claim, Cadbury Schweppes PLC, Plano, Tex., recently reformulated and remarketed its 7Up beverage as a five-ingredient, "natural" product.

Get Well Soon

The twin epidemics of obesity and diabetes dominate the health and wellness category. No day passes without the mention of one, the other or both on television, radio or in newspapers. But in general, between one-fourth and one-third of consumers make food choices based on health for some reason.

This is a trend that plays directly to our desire to ingest specific foods or beverages for the purpose of preventing or palliating a disease or condition. The reason antioxidants, botanical extracts and the whole foods (berries, teas, soy) that contain them are critical underpinnings of this trend is because of the promise such items hold to improve how you feel and perform.

From an ingredient standpoint, health and wellness concerns offer the best variety of options for processors. A manufacturer developing a product in this market has literally thousands of botanical extracts, antioxidants, phytochemicals, carbohydrate compounds (such as sugars, starches and fibers), protein compounds or fractions and healthy oils from which to choose. The trend to address health reached its mainstream "tipping point" in food processing with the decision in 2004 by America's second largest cereal maker, General Mills Inc., to reformulate all of its breakfast cereals to be based on whole grains. There certainly were defining moments along the way – fortification of flour and breads with folate, calcium enrichment of juice and other "health-value added" movements all are good examples. But the Minneapolis- based company was the first processing giant to change its entire line of products in such a manner.

Meanwhile, emphasis is shifting away from dieting and related fat and calorie-count issues. In a survey by Mintel International, Chicago although seven in 10 American adults claim to be trying to eat healthier foods, almost that many – 65 percent – say calories don't always count, with about half of Americans finding nutritional value the important factor.

According to NPD Group Inc., a research firm in Port Washington, N.Y., obesity rates have held steady for four years now. Diet still will be big; as are more than 60 percent of Americans. But our interests will turn toward other health concerns. Allergies (including gluten and lactose intolerance), energy, immunity and the more general "feeling better" issues are moving up to occupy a growing portion of the megatrend.

Another study by Mintel notes about 36 million Americans claim to suffer from either a food allergy or intolerance. And a Natural Marketing Institute report finds two-thirds of baby boomers are most afraid of fatigue as they age, and nearly half are worried about diminished mental capacity.

From a food and beverage manufacturing point of view, all these wellness areas are showing some of the strongest growth potential.

Modern Biotechnology in Food: Modern Biotechnology and Food Quality

In Europe, a vast diversity of high quality foods provide the carbohydrates, fats, proteins, minerals and vitamins needed in the everyday diet of consumers. At the heart of food production is biotechnology. One aspect of biotechnology which has been used for centuries is the selective breeding of crop plants and farm animals to produce improved food. Another is fermentation, in use for millennia to produce fermented foods like cheese, bread, beer, sauerkraut and sausages. The first use of gene technology two decades ago opened up the potential for many additional advances in both selective breeding and fermentation. Each specific step forward might be relatively small, but together they could add up to further improvements in the nutritional quality, appearance, flavour, convenience, cost and safety of foods.

Better Raw Materials

In improving raw food materials, many plant breeding programmes have been directed towards boosting yield or allowing more environmentally compatible agriculture by increasing the resistance of crops to viruses, pests or herbicides. Increasing yield has clear benefits in helping to feed the world's ever-increasing population and could provide cheaper food.

Plants which are resistant to attack by insect pests and diseases would need fewer pesticide applications; resistant crops such as maize, tomatoes and potatoes are already being developed. Crops have also been produced with tolerance to modern, more environmentally compatible herbicides, with the aim of achieving optimal weed control with reduced levels of herbicide. Today, there is increasing interest in improving the nutritional value, flavour and texture of raw materials. This could help encourage greater fruit and vegetable consumption in line with government guidelines on healthy nutrition.

A range of promising crop plants are being developed with:

- Improved nutritional value - Crops in development include soybeans with a higher protein content; potatoes with more nutritionally available starch and with an improved amino acid content; pulses such as beans which have been altered to produce essential amino acids; crops which produce beta-carotene, a precursor of vitamin A; and crop plants with a modified fatty acid profile. An example is a strain of oilseed rape which produces a special type of polyunsaturated fatty acid (the so-called w3-fatty acids). These have been linked to brain development and have potential in a range of speciality, clinical and infant foods.
- Better flavour - For example, types of peppers and melons with improved flavour are currently in field trials. Flavour can also be improved by enhancing the activity of plant enzymes which transform aroma precursors into flavouring compounds.
- Improved keeping properties with the aim of making transport of fresh produce easier, giving consumers access to nutritionally valuable whole foods and preventing decay, damage and loss of nutrients. Examples include the improved tomatoes now being sold in the US, and recently approved in the UK, which have been genetically altered to delay softening. Research is

underway on making similar modifications to broccoli, celery, carrots, melon and raspberries. The shelf-life of some processed foods such as peanuts has also been improved by using raw materials with a modified fatty acid profile.

- Reduced levels of toxicants, allowing a wider range of plants to be used as food crops, such as the edible strain of sweet lupin which has been developed through conventional breeding techniques.

Improved Food Ingredients

Necessary changes to the key food ingredients, starches and oils, are usually made by processing. Biotechnology opens up the possibility of altering crop plants to produce exactly the type of ingredients needed:

- Starches- Plant breeders have introduced a bacterial gene into potato plants which increases the proportion of starch in the tubers whilst reducing their water content. This means that the potatoes absorb less fat during frying, giving low-fat chips. Sweeter potatoes have also been produced which have a higher sucrose content than traditional varieties.
- Oils- Both rapeseed and sunflower are being altered to produce more stable and nutritious oils, which contain linoleic acid instead of linolenic acid and have a lower saturated fat content. Rapeseed has also been modified to produce a high-temperature frying oil low in saturated fat.

Advances in Processing and Additives

Research underway at present aims to allow the production of better food raw materials by crop plants. However, some processing steps remain essential to bridge the gap between currently-available raw materials and the desired end-product.

Traditional biotechnology has played a major role in producing fermented foods- where desirable changes are produced by the action of micro-organisms or enzymes- of which over 3,500 different types exist around the world.

In Europe and North America, bread, yoghurt and cheese are perhaps most familiar. In Africa, foods made from fermented starch crops like yams and cassava are more important, whereas in Asia, products derived from fermented soya beans or fish predominate.

Fermentation can make the food more nutritious, tastier or easier to digest, and it can enhance food safety. It also helps to preserve food and to increase its shelf-life, reducing the need for additives. Genetically improved strains of microbes can make a major contribution to these desirable properties.

For many years, a wide range of additives, processing aids and supplements have been obtained from microbial sources by fermentation. Increasingly, modern biotechnology is being used here. Products include vitamins, citric acid, natural colourings, flavourings, gums and enzymes. Gums used as low-calorie thickening agents and low-calorie sweeteners from natural ingredients are also produced using modern biotechnology. Enzymes- the naturally-occurring catalysts responsible for literally all the biochemical processes of life - are used in applications such as bakery and cheese making to improve texture, appearance and nutritional value, and to generate desirable flavours and aromas.

A second area where biotechnology has advantages is in improving the processes by which food is produced. Here, it can be used to develop mild, highly specific processes using modified micro-organisms and purer, cheaper enzyme products. These can offer better productivity, cost-effectiveness and energy efficiency than existing processes. They can produce top-quality foods with a reduced need for additives such as flavourings, and can also reduce the environmental impact of food processing.

Specific areas of food processing where advances are being seen are:

- Bread-making, for which improved strains of yeast have been developed containing genes for production of other food processing aids, such as amylases, which give improved dough. Yeast can also be used to produce a range of enzymes for use in processes such as cheese production, where introducing a copy of a calf gene has given a strain of yeast which produces the enzyme, chymosin. Previously, this enzyme could only be obtained from the stomachs of calves.
- Fruit juice production, where juice yields from apples can be improved by adding pectinase enzymes. These are produced naturally by a strain of the mould Aspergillus. The rate at which the enzymes are made can be improved by transferring the gene for pectinase from one strain of the mould into a second strain with a higher capacity for enzyme production.
- Improved quality management and food safety, through a greater understanding of micro-organisms and enzymes in food production. A range of biological tools, such as monoclonal and

polyclonal antibodies, will add to this impact through their use in a range of diagnostic tests aimed at enhancing the quality and safety of products and processes. These can potentially be used to monitor the presence of additives, toxins, pesticides, micro-organisms and antibiotics, and they will give quicker, more accurate detection than traditional laboratory processes.

Health is still the biggest part of the aging trend. For every age group there's a health concern some processor is targeting. Attention is split mostly among concerns of children and seniors. By sheer numbers, the last group is headed for steady growth as a trend. Basically, like Dylan Thomas, we will "not go gentle into that good night." Besides, 60 is the new 40, mirrors be damned. If a food or beverage can boast an ingredient polyphenolics, antioxidants, omega oils – to put some bloom back on the rose, then we're going to buy it. But the aging trend involves more than ingredients and formulations. The doubling of the over-65 population by 2030 means increased need for easier-to-open containers. The logistics involved can include everything from packaging machinery redesigns to food safety concerns based on conflicting needs to creating tamper-resistant and sanitary packaging that doesn't require sophisticated kitchen tools or a magician's touch to open.

Then there's labelling. Larger print on labels for presbyopic eyes means less room for those marketing blurbs, images, slogans and serving suggestions that go so far to separate one product from its competitors.

Eat Global, Buy Local

New findings in a joint study by Mintel and the National Assn. of the Speciality Food Trade, New York, show speciality food "continues to show strong mainstream movement," and it singles out ethnic influences as part of the growth surge in the $35 billion product niche.

"American consumers are continuing to take advantage of the country's diverse cultures and offerings," said Marcia Mogelonsky, senior analyst for Mintel. "As consumer interest in new flavours and products continues to grow, so does the speciality food market. As immigrants continue to acculturate, their food traditions are becoming more mainstream. More Vietnamese, Thai and Indian flavours will continue to flourish within this category."

Globalization can be a volatile category. One example is the healthy fruit açai. Before 2000, the fruit was enjoyed primarily on its South American home turf, usually at small juice stands. In 2001, the first company to bring açai to the U.S., Sambazon, San Clemente, Calif., imported 40,000 lb. of açai pulp. In 2005, the figure was about 4 million lb., according to CEO Ryan Black, and demand is growing as fast as supply can fill it.

Globalization's real heft as a trend, though, comes through such mega issues as the drive to enter the China market. (To find out more, check out our archived web cast, "Four steps to China: One billion hungry customers await you"). And as the sleeping tiger that is India awakens, another billion potential customers wait in the wings.

Control Yourself

We're controlling portions not just for health but convenience. As a trend, convenience has been high on the list of movements to follow for years. But the two aspects merged in 2004 when Kraft Foods Inc.'s Nabisco brand launched 100-Calorie Packs of some of its most popular cookies and crackers. Those aren't exactly health foods. But it was a clever move, because dietitians and other nutrition experts had been clamoring about portion control for years. Although aimed largely at the restaurant industry, it was a pleasant surprise when the cry for sensible portions finally was heard by the packaged food industry.

The trend took off with such gusto Kellogg Co., Battle Creek, Mich., brought out its 90-calorie packs of Special K Snack Bites and 130-calorie Granola Munch'ems. Frito-Lay North America, Plano, Texas, joined in with 100 Calorie Mini Bites versions of its Cheetos and Doritos snacks. In the space of two short years, sales of such portioned packets passed the quarter-billion dollar mark.

Even manufacturers of healthy foods are hopping on the portion-control bandwagon. For example, Nspired Natural Foods Inc., San Leandro, Calif., just went national with its 90-calorie O'Coco's chocolate crisp packs. Vitalicious Inc., New York, offers 100-calorie, natural, vitamin-fortified brownies and muffins.

Make Room Kosher, Halal is Here

Kosher broke away from ethnic as a trend of its own with the first wave of fear over mad cow disease. Halal certification, the Muslim equivalent of kosher, is finally grabbing at the same brass ring.

The religious oversight of food encompasses food safety, health and wellness, ethnicity and spiritualism with one stamp. Or two. In the past year or so, the growth of religious adherence among boomers is seeing the combining of kosher and organic. This trendlet ("trendmerge?") is still too new to show reliable growth figures.

However, companies such as Wise Kosher Natural Poultry Inc., Brooklyn, N.Y., are literally betting the farm on it.

Kosher/halal isn't just about meat, although the timing couldn't be better with meat sales jumping. The number of products with some sort of kosher certification is nearing six figures. Estimates are that three-fourths of manufactured foods and beverages have, are in the process of, or are seeking some sort of religious oversight certification.

In a 2005 Kosher Food Report by Mintel, it was reported that 21 percent of food purchasers knowingly buy some kosher products, and 28 percent of all consumers purchasing kosher products are driven by taste and quality. Consumers, it was noted, see kosher as a synonym for quality. Consumers see such products as being "safer," "healthier" and "better for you." Vegetarians have learned to trust the pareve (neither meat nor dairy) designation on kosher products because of the strictness that permits not even a trace of dairy via ingredient, equipment or handling.

In addition to vegetarians, Hindus, Muslims, Seventh-Day Adventists and just generally watchful consumers go for kosher. Sales of these certified products top $100 billion annually and are projected to increase 14 percent in 2007.

The half-trillion-dollar food and beverage business is trend-driven, make no mistake. The value to processors is in growing product lines not only to serve but to steer these trends. This doesn't mean every new product needs to be an organic, whole-grain, 100-calorie, ethnic-oriented and kosher snack loaded with omega 3s and anthocyanins with an easy-open top and large-print labels. But indications are such a product would sell well. As long as we remember that at the end of the game, taste trumps all trends.

Table: *Top trends in food processing (and their trendlets)*

Trend	***Trendlets***
1. Organic	Non-GMO, Fair Trade, Sustainability, Regional, Minimalism, "Natural"
2. Health and Wellness	Diabetes and Obesity, Kids' Health, Food Safety, Women's Health, Allergies and Immunity, Well-Being, Energy
3. Age Awareness	Aging, Teens, Kids
4. Portion Control	Serving Size, Convenience
5. Globalization	Ethnic Flavours, Multinational Production Regulations
6. Kosher/Halal	Food Safety, Certification and Oversight, Spiritualism

Beverages Take Two Paths

A major trendlet is bubbling up in the world of potables. Beverages have become more or less. That's not a misprint. With sales of conventional soft drinks dipping for the first time in years, the things we drink are polarizing into quaffs with "more" (concentrated meal-replacements, energy drinks and smoothies) and "less" (flavoured and enhanced waters substituting nutraceuticals and exotic fruit extracts for calories). Manufacturers with examples in the first category include such companies as Unilever/Slim-Fast, Odwalla, Naked Juice and Soyblendz. In the latter category are such no- and low-calorie refreshers as Glaceáu Fruitwater, Gus Grown-up Soda and O2Go. Tea is big again, too. Thanks to green tea being recognized for its health benefits, the category is now flooding the market with tea and tea-fruit juice blends that lean toward exotic fruits to take advantage of the health and wellness trends as well as globalization's growing ethnic flavour offerings. Such combos are including tres chic fruits like açai, carambola and pomegranate.

Technological Trends and Needs

Several well-established traditional processing options are available for the preservation of food. The most widely used among them, thermal processing, provides a high degree of microbial safety. It tends, however, to degrade the quality of foods to some extent. Freezing and distribution of frozen foods is an important technology which retains very well the nutritive quality, but changes the physical state and consumes a great amount of energy. The present is a time of rapid change in the field of food technologies and the pace of change is increasing. Major motivations determining trends of development of new, emerging or future food technologies are those which signify responses of food science and industry to:

- demands of consumers due to their changing lifestyles and expectations for
- fresher, more natural foods, which
- are less severely processed (less heat- or freeze-damaged),
- contain less preservatives, or are even free from "artificial" additives;
- nutritionally more advantageous food (e.g. containing less salt, less sugar or fat);
- safer food (posing no microbiological or chemical health hazards); and at the same time are

- foods convenient to handle ([semi] prepared or ready for consumption and with a sufficient shelf-life);
- needs for less energy requirement of processing;
- the necessity for lower impact on the environment.

These - to some extent conflicting - requirements motivated the introduction of less severe or 'minimal processing' technologies such as controlled atmosphere storage of fruits and vegetables, and modified atmosphere packaging of foods, or the development of extended shelf-life refrigerated foods such as 'sous-vide' cooked products. These requirements have also resulted in growing interest in new, 'non-thermal' methods of food preservation such as

- using new 'biopreservatives' or new 'protective cultures' utilizing their antagonism against pathogenic microorganisms;
- new 'physical' technologies:
  — ionizing radiation treatment;
  — high hydrostatic pressure treatment;
  — high voltage electric field pulses.

To inactivate pathogenic and spoilage microorganisms. (Other new physical methods of antimicrobial treatment such as high intensity light pulses, "manothermosonication" (combination of pressure, heat and ultrasound), or treatment with oscillating magnetic fields (FARKAS, 1997; GOULD, 2000) are either of more limited scope, or they are not yet sufficiently ready scientifically or technically for implementation). Application of the technologies listed above offers various opportunities for mildly processed products by preserving their sensory quality, nutritional value and appearance. However, the application potential of any new technologies, which are coming from research laboratories and not 'sanctioned' by centuries of empirical use, is influenced by many factors:

- technological feasibility;
- technical possibility;
- health impact
- wholesomeness of the product;
- occupational safety;
- environmental friendliness;
- economic feasibility (including their energy demand);
- infrastructural conditions/requirements;
- investment need and availability of investment power;

- political attitude;
- social consequences;
- psychological aspects/risk-benefit perception.

Whereas the first four aspects listed can be scientifically studied, and clear-cut general answers can be given according to the status of science and technology of a given epoch, the other factors are very much interrelated and depend on local conditions.

A brief sketch of the state of the art of the new/emerging technologies can be given as follows:

> *"New" natural antimicrobial substances, particularly the use of bacteriocins produced by some lactic acid bacteria seem to gain a role in eliminating the risk of specific microorganisms in some foods. Some strains of lactic acid bacteria are considered as protective cultures, if they can be inoculated into certain foods such as vacuum-packaged processed meats, and assert an inhibitory effect on pathogens during storage at abuse temperature, while having negligible effect on the sensory quality of the products. Because of associated flavours that can alter the taste of food, future uses of plant-derived antimicrobials as food preservatives are not likely on their own, but as part of a preservation system.*

Ionising radiation is a versatile form of processing energy used already in a wide range of non-food applications. It offers various technological benefits by reducing food losses and improving food safety. Irradiation extends to solid and semi-solid foods like meat, poultry and seafood the same benefits as thermal pasteurization provided for liquids. Radiation treatment at doses of 1.5 to 7 kGy - depending on conditions of irradiation and of the food - can effectively eliminate potentially pathogenic non-sporeforming bacteria from suspected food products without affecting nutritional and technical qualities. These bacteria include both long-time recognized hazards such as *Salmonella* spp. and *Staphylococcus aureus,* as well as emerging or "new" pathogens such as *Campylobacter* spp., *Listeria monocytogenes* or *Escherichia coli* O157:H7. In addition to control of the aforementioned bacteriological hazards, "radicidation" of perishable commodities can extend their shelf-life 2- to 3-fold, and inactivate food-borne parasites, the latter being of particular importance for developing countries.

After decades of unprecedented intense and wide-ranging research efforts, food irradiation is now a well understood and controllable food processing technology supported by all relevant specialized agencies of the United Nations (the World Health Organization, the Food and Agriculture Organization and the International Atomic Energy Agency) and many national scientifically authoritative bodies in different countries. Its slow implementation can be explained by the long time, which was needed to demonstrate adequately the safety and wholesomeness of irradiated food, the lack of readily available radiation facilities and their investment-demanding character, as well as an inadequate awareness of problems, which justify the use of this technology.

The safety and nutritional adequacy of irradiated food have been well established, the technology is ready to use technically, and the need to improve the microbial and parasitic safety of food became a major driving force behind the implementation of food irradiation in both developed and developing countries. When irradiation is perceived as an adjunct to, but not a replacement for, GMP, it can serve as a 'critical control point' in the Hazard Analysis Critical Control Points system of safety management, a concept becoming mandatory in more and more countries.

The advantage of *high hydrostatic pressure* (HP) treatment (up to 900 MPa [i.e. as high as 9000-times the atmospheric pressure] for several minutes) is that it treats all parts of a high moisture food equally (high isostatic pressure) and it is attractive from the point of view of food product quality. At present its application is limited to a pasteurization-like process for certain foods in which bacterial spores are not a problem (low-pH products) or which have a limited chill shelf-life. Success of its further development depends on effective control by appropriate combination treatments and additional research for careful establishment of performance criteria for reduction of the number of relevant pathogens by a required safety factor (GOULD, 2000). Due to these facts, and because of costs involved in HP pasteurization/ sterilization of foodstuffs, this process will remain probably rather specialised and will only be used to preserve foods for which a premium price can be obtained.

High voltage electric field pulses (HELP) to effect non-thermal inactivation of microorganisms in foods was also explored, and these studies led to the development of prototype and industrial scale devices recently. HELP treatment is the application of pulses of very high field strength (2-5 $V.um^{-1}$) for a very short time (microseconds) to foods between two electrodes. The treatment requires fairly complex electronic and fluid handling systems. Studies on combination processes have shown potentially useful synergies because "electroporated" bacterial cells become much more sensitive than untreated cells (e.g. to bacteriocins). The application of HELP will probably be limited to liquid foods or liquids containing small particulates. There are still, however, considerable knowledge gaps that will need to be addressed, and regulatory hurdles to be overcome, before commercialisation of the technology, but HELP products might be in the market place in 10 years' time.

All these technologies might be utilized even more efficiently in rational combinations with other preservative treatment(s). Among future challenges for food science in the field of more efficient improvement of microbial safety and quality maintenance of food are:

- better understanding of factors affecting the establishment, survival and growth of food-borne microorganisms by multidisciplinary research on
  - — physiology of microorganisms;
  - — development of predictive models for their growth, survival, inactivation or the shelf-life of foods;
  - — effect of food structure on its interactions with microorganisms;
  - — application of molecular biology for improved characterization, typing and detection of important microorganisms;
- better utilization of synergistic interactions of "combinations"/ "hurdle effects", but avoiding stress adaptation.

Advances in predictive microbiology will assist optimization of existing processes and formulation of alternative processes with regard to their effect on food safety. If the microbiological feasibility of new physical technologies is carefully established and, particularly, new combination approaches are developed and proven effective, the opportunities for the use of some of the new techniques are likely to expand in the future. From the food industry side, there is a need to "metabolise" the novelty of the aforementioned technologies to appreciate their potential over the hurdles that must be overcome.

Food with Improved Nutritional Value

Until recently, health aspects of food were mainly thought of as the absence of detrimental components, in particular additives and unwanted compounds. In the past years, however, it has become evident that some naturally occurring components of certain foods can help to maintain a state of well being and health through optimising our body functions or to reduce risk of chronic diseases, such as certain cancers and coronary diseases. This is particularly important in that most countries are experiencing a considerable rise in the proportion of elderly people, and proper nutrition can assist us to live not only longer, but in better health.

In addition, many technological innovations are related to the recognition of the important role of food in promoting and sustaining health, resulting in the concept of "*nutraceuticals*" and "*functional foods*". Nutraceuticals are those food components, which play particular roles in maintaining health. They originate mainly from plants, i.e. they are "*phytochemicals*". Functional foods are foods characterized by the presence of one or more components having beneficial physiological effect and being effective in the maintenance of good health. In one group of functional foods live bacteria (certain lactobacilli and bifidobacteria) provide the main beneficial effect. These foods are called *probiotics*.

Other functional foods contain *prebiotics,* i.e. substances that facilitate growth of the beneficial bacteria in the intestinal tract of the consumer. This interaction between food, nutrition and health is a new challenge for both food science and the food industry and it represents a positive approach to 'optimizing" nutrition. It is important, however, that claims on health-promoting effects have sufficient scientific substantiation.

It is important that health authorities should establish criteria for mandatory qualification of functional foods. Development of functional food should not be only a marketing claim, but rather a scientific challenge. The presence of a bioactive compound in a food does not necessarily ensure that it will be biologically active when it is consumed. The benefits of the "functional foods momentum" will not be realized unless scientifically sound and non-misleading messages are provided to consumers.

Food can be said to be "functional" if it is satisfactorily demonstrated to affect beneficially one or more target functions in the body beyond adequate nutritional effects, in a way which is relevant to either the state of well-being and health or the reduction of the risk of a disease. A functional food can be a natural food, a 'modified' traditional food or a 'novel food' as defined in the aforementioned EU Regulation.

The production of functional foods happens by:

- eliminating a component causing deleterious effect to the consumer (e.g. an allergenic protein);

- increasing the concentration of a natural component which induces beneficial effects;
- adding a *natural* substance which is not normally present, but for which beneficial effects have been demonstrated;
- replacing a component which causes deleterious effects by a component with beneficial effects;
- improving the bioavailability of food components with beneficial effects.

Functional foods must remain foods, they are not medicaments, but part of nutrition/diet. Because functional food development targets healthy subjects, proving their effects' statistical significance requires a special methodological approach. It will become increasingly important to:

- identify the mechanism of action of the active components in functional foods;
- clarify the impact that commercial processing and home-processing have on bioactive components within foods;
- optimize process parameters for maximal retention and increased bioavailability of beneficial compounds;
- understand the dynamic interactions that occur among the various components of not only the food but with the other constituents of the entire diet;
- investigate the interaction between diet and human intestinal microflora and its implication for health.

Transgenic Food

While consumers accept the functional food concept remarkably readily, the acceptance of novel foods as they are defined according to the EU Regulation (EC, 1997), and particularly transgenic food, is controversial.

Conventional breeding can be used to transfer genes only between sexually compatible organisms. Complementary to conventional breeding techniques, gene technology allows the transfer of genes between unrelated species. Thereby, breeding targets can also be achieved more quickly both in plant and animal breeding. This is one of the key, but hotly debated technologies of our times. Important topics that need to be addressed vary from legislation, such as labelling requirements, to safety and environmental issues. Concerns about application of this agricultural biotechnology are on the ecological impact of growing genetically modified foods, the impact of these crops on biological diversity, and on the safety of food supply, or the development of resistance by insect pests. However, the potential of the agricultural new biotechnologies is enormous also for developing countries. Therefore, questions about agricultural biotechnology must be addressed for people in both developed and developing countries, as we have to address the issue of food security for a world population of some 9000 million people in the year 2050. Furthermore, genetic engineering is not just a new technology for crop improvement, it is a powerful research tool that is helping to provide fresh and better insights of molecular mechanisms involved in biological processes.

It is forecasted that in the next decade about four dozens of agricultural crops will be genetically modified. It is estimated that already 8.3 million ha of genetically modified corn (20 % of the total cultivated area) were planted in the US in 1999, and that more than 66 million ha will be used to cultivate transgenic plants by 2005 in North America. No less remarkable are the opportunities and research results in the field of food biotechnology by improving microorganisms used in food fermentations, and on the exploitation of microorganisms for the manufacture of food ingredients.

Without venturing more deeply into this enormously complex problem-area where I have no expertise, I should like to limit myself to those aspects of transgenic food or food components which are relevant to the potential improvement of nutritional value and technological functionality of food or food components.

There are already a number of foods on the market which are produced using genetically modified organisms or containing GM ingredients e.g. chymosin used in cheese making, use of GM tomatoes in paste, and GM soybean and corn products. Thus, most of the transgenic products to date have been developed for agricultural and processing efficiency, and not yet with direct consumer benefits in mind such as improved taste and higher nutritional value. I share the views of those who consider that a "second generation" of transgenic crops should be devoted to achieve these advantageous compositional changes in order to serve the increased interest in functional foods and passing the benefit on to the consumer. Food R&D in the 21st century should also support consumer oriented product development because it has the potential to become part of the health care system. Several transgenic foods are now under development, which aim to develop properties of nutritional significance. Some examples:

- producing rape seed and corn with nutritionally more favourable oil composition (containing negligible levels of erucic acid);

- spinach and lettuce accumulating less nitrate;
- potato richer in starch, thus absorbing less fat when fried;
- certain cereals with increased lysine content;
- certain legumes with increased methionine content;
- rice not producing an allergen;
- "yellow rice" capable of synthesising beta-carotene;
- strawberry and broccoli producing higher levels of anticancer and antioxidant agents.

Safety evaluation, however, is a key issue, which must be addressed in relation to the development of novel or improved foods. Guidance on this complex topic is available from a report of a Joint FAO/WHO Consultation (1996) on "Biotechnology and Food Safety" (FAO/WHO, 1996).

Regarding safety assessment of novel food in relation to nutrition, special attention must be paid also to allergenicity of those foods which are produced using this modern biotechnology to avoid the potential appearance of a major food allergen in a product that is normally allergen free.

A summary of results of an EU project on the development of new methods for safety evaluation of transgenic food crops has been presented at the 3rd Karlsruhe Nutrition Symposium in October 1998, while the International Life Science Institute recently published "Consensus Guidelines" on the safety assessment of viable genetically modified microorganisms used in food (ILSI, 1999).

Principles for approving novel food, particularly genetically modified raw materials, were fundamentally different in North America than in the European Union, which created complications for the food industry and trade. The labelling of genetically modified organisms was not mandatory in the US, where the FDA labelling policy requires "biotech foods" to be labelled only if they are significantly altered. In response to a request from FDA for comments from the public on the Agency's current regulatory regime for food biotechnology and labelling, in formal comments submitted to FDA, the National Food Processors Association (NFPA) called, early in the year 2000, for a compulsory notification process prior to the marketing of new "biotech" food products. The producers of such foods should file with FDA summary documentation to support the determination of safety for the biotech food (NFPA, 2000). NFPA further supports the use of voluntary labelling of foods to indicate the presence or absence of bioengineered ingredients.

The European Union "Novel Food Regulation" established a system for formal, mandatory pre-market evaluation and approval for most innovative foods and food production processes, placing particular emphasis on genetically modified products. It requires additional specific labelling of "any characteristic or food property such as composition, nutritional value or nutritional effects, or intended use which renders the food no longer equivalent to its conventional counterpart." Although this EU Regulation provides broad guidance, it leaves the door wide open to interpretation and it is unlikely to stimulate European innovation and competitiveness. The recently finalised "White Paper on Food Safety" of the Commission of the European Communities states that "the Community provisions governing *novel foods* have to be tightened and streamlined", and it describes actions to this end (EC, 2000).

This is an area where the future depends on acceptance of rational, science-based weighting of risk with benefits and the provision of accurate and unbiased information to the consumer. The risk of not using technologies and of alternative methods should be also concerned. However, there is a multitude of ethical and social issues to be considered. There appears to be an even greater aversion among consumers towards the genetic engineering of animals than towards plant biotechnological programmes.

To analyse the risks of transgenic foods, the FAO/WHO Codex Alimentarius Commission established an "Ad Hoc Intergovernmental Task Force on Foods derived from Biotechnology" (FAO, 2000), which held its first meeting in Japan at the time when I was preparing this paper. This Task Force uses the concept of "substantial equivalence" established by the Organization for Economic Cooperation and Development as a central to the process of risk assessment and will review other methods for science-based risk assessment. It is expected that the above Codex Task Force will come up with a Codex' standard on transgenic foods. The "precautionary approach" adopted by the recently signed UN Protocol on Biosafety (Cartagena Protocol) can be used by governments in their risk management relevant to the above risk assessment. Identifying and labelling foods as having been derived from biotechnology can form a risk management system that will both protect the health of consumers and promote fair trade practices (FAO, 2000).

In my view, trusting science is one of the key factors for developing new technologies and evaluating new products, regardless of the technology or practices concerned. Unfortunately, however, food irradiation, biotechnology, and confidence in food safety are all media-vulnerable issues. Future developments will depend on more effective and balanced communication to the public, effectively addressing social, ethical and political issues, scientific questions and regulatory needs. It is important to understand the way that people perceive risk psychologically. Consumer acceptance of new processes is likely to be increased when a direct consumer benefit is recognized. However, huge information gaps exist among scientists and particularly between scientists and consumers. To match the promise offered by technological advances and optimize nutrition, the overcoming of barriers in psychological and - as a consequence - political feasibility is required. This needs not only research efforts but education in all stages of the food chain and in all sectors of communication, because "if we fail to train we fail to convince, if we fail to convince we fail".

Nutritional Quality

Food Composition

- *Energy* for the metabolic and physiological functions of humans is derived from the chemical energy bound in food and its macronutrient constituents, i.e. *carbohydrates*, *fats*, *proteins* and *alcohol*, which act as substrates or fuels. After food is ingested, its chemical energy is released and converted into heat, mechanical and other forms of energy
- Food is made of one or several *macronutrients*, which are carbohydrates, fats, proteins and alcohol (e.g. rice contains 0.5% fat, 7% proteins, and 80% carbohydrates). Food contains also *micronutrients*, such as minerals (e.g. iron, calcium, etc.) and vitamins (e.g. vitamin A, B1 or C).
- Other components of the food are *water*, and sometimes *antinutritional factors.*
- Values for nutrients in food can *vary widely* and depend on such things as: the growing environment, the particular variety or breed, the method of processing, the storage temperature and time etc. The values given in the food specifications section are an indication of nutritional value.

Energy Provided by Macronutrients

Energy is needed for essential body functions (such as breathing), growth (especially during childhood), and physical activities (working and playing).

Macronutrients provide different amounts of energy, expressed as kilocalories (Kcals). Fat provides approximately twice as much energy (9 kcals/g) as the same weight of protein or carbohydrate (4 kcal/g). As stated above, more carbohydrate than fat is usually eaten in developing countries and, therefore, most food energy in the diet in these countries is derived from carbohydrate sources. The relative concentration of protein and fat in the diet is important and is expressed by the percentage of energy in the diet provided by either fat or protein. For example, if a diet provides 2,000 kcal, of which 200 kcal is provided by fat, the fat-energy percentage is 10.

Energy and Protein Requirements

The total amount of energy and protein needed by different individuals varies a great deal, depending primarily on the amount of physical activity but also on age, sex, body size and, to some extent, climate. Extra energy is needed during pregnancy and lactation. An average population, made from people of all age and gender, will need 2,100 kcal per day. Energy requirement will vary with age, sex, physical activity, climate, etc.

Energy Density Concept

The quantity of energy that a child can consume each day from gruels depends on the number of meals, quantities consumed during each meal and the energy density of the gruels. However, in many societies, the mothers, involved in multiple tasks, cannot prepare gruels more than twice a day. In addition, babies cannot eat more than 30 to 40 ml of gruel per kilogram of body weight during each meal because of their reduced stomach capacity. In the case of gruels prepared from starch-based foods that have not undergone enzymatic or hydrothermal treatments (treatments involving water and temperature such as extrusion-cooking, drum-drying), the concentration of flour in the gruels is the main determinant of their energy density. The first approximation is that 4 kcal can be obtained from 1g of dry matter.

These gruels have a viscosity which increases very quickly according to their concentration in dry matter. Those who prepare these gruels are thus placed before the following dilemma: increase the proportion of flour with respect to water and obtain a gruel of very high viscosity, *difficult for children to swallow*, or prepare gruels of suitable consistency but of low energy density, and thus give more than three meals a day. There is only one way to increase the quantities of energy consumed by children, wherever gruels are not given more than three times per day, only one way: increase the energy density of the gruels. For that, the flours must undergo enzymatic and/or hydrothermal treatments which modify the physicochemical properties of the starch. These treatments break the starch macromolecules, limit their swelling during cooking, and consequently change the viscosity of the gruels. It thus becomes possible to prepare gruels of higher energy density while maintaining an appropriate consistency.

Measurement of Gruels/Porridges Consistency

When a mother prepares a gruel for her child, it is undoubtedly the desired consistency of the gruel which guides her in the choice of the flour/water proportions. However, from these proportions depends the *energy density* of the gruel and, consequently, its' *nutritional value (i.e. its nutrient density).* Thus, the higher the relative quantity of flour, the more the gruel will be nutritious, but also the more will it be thick, until the maximum consistency threshold accepted by the mother or the child is reached. The manufacture of infant flours permitting the preparation of gruels of energy density likely to satisfy the energy needs of the child, necessitates the application of technologies that reduce viscosity: addition of amylolytic enzymes, malting of a fraction of the raw materials, *extrusion-cooking*, *drum-drying*, etc...

In order to study the effectiveness of these technological treatments, an evaluation of consistency is necessary. The parameter generally measured is *viscosity*. But out of the laboratory, the realisation of reliable viscosity measurements is very difficult. In addition, the absence of recommendations for the adoption of standardised conditions of measurement invokes a great disparity in the measurement methods employed. There are many viscometers of different trade marks and characteristics and from one instrument to another, the measured viscosity on the same gruel can vary up to a factor of 10.

To evaluate the consistency of gruels, it is also possible to use a *Bostwick* consistometer. This empirical measurement, relatively simple to use in an enterprise, is less well conceived than the measurement of viscosity but could be better correlated with the sensory appreciation of mothers. Research studies are currently being carried out in the Laboratory for Tropical Nutrition, IRD, in order to better understand the relationship between viscosity, *Bostwick* flow and the sensory appreciation of the consistency of gruels.

The compartment that has a guillotine-type system is filled with a given volume (100 ml) of gruel. At t=0, the gruel is released and the consistency parameter obtained corresponds to the distance covered by the gruel front after 30 seconds of flow. The only control condition of the measurement is the gruel temperature: generally, a temperature close to that at which the gruel is consumed, that is to say approximately 45 °C is adopted.

Macronutrients

Protein

Proteins are made up of 'building blocks' called *amino acids*, composed of carbon, hydrogen, oxygen and nitrogen (amino group). Proteins from different food sources contain different amounts of amino acids. Proteins from animal origin, such as meat, milk and eggs, contain all *essential amino acids* in balanced amounts. Essential amino acids are those that the body cannot make itself and must therefore be eaten. In contrast, proteins of vegetable origin (e.g., cereals and pulses) contain on their own insufficient quantities of some of the essential amino acids. By combining different foods, however (e.g., cereals with beans), adequate levels of all amino acids can be obtained without requiring protein from animal sources.

Proteins are required to build new tissue, particularly during the rapid growth period of infancy and early childhood, during pregnancy and nursing, and after infections or injuries. Excess protein is burned for energy.

The nutritional value of a protein food can be judged by its ability to provide both the quantity and number of essential amino acids needed by the body. Different food sources contain different groups of proteins, which are made up of different arrangements and amounts of amino acids. In general, proteins from animal sources are of greater nutritional value because they usually contain all the essential amino acids.

Proteins from plant sources, such as cereals and vegetables, may be deficient in one or other of the essential amino acids. For example, the proteins obtained from wheat lack adequate quantities of one essential amino acid, and those from beans are deficient in another but the combiantion of cereal and pulses will provide a balanced diet. Because the content of amino acids is different in each food, when they are eaten together they complement each other and the mixture is of higher nutritional value than the separate foods, and is as good as animal protein. It is important, that a variety of different types of protein foods are eaten.

Cooking can alter the amino-acid composition of protein and this usually results in desirable flavour and browning development. Very little nutritional value is lost.

Protein Intake

WFP recommends that 10 - 12% of the energy of the ration is derived from protein.

For instance for a daily ration of 2,100 kcal;

- 10% = 210 kcal
- 210 kcal/4 kcal/g = 52.5g of protein per day
- Maize contains 8.5g of protein/100g.
- Beans (Cowpeas) contain 23.5g of protein/100g.
- A ration made from 400g of maize (400 x 8.5% = 34g) and 100g of Cowpeas (100 x 23.5% = 23.5g) will bring 57.5g of protein.

Fat

Fat is also known as lipid and is mainly present in food in a form called 'triglycerides'. Triglycerides consist of glycerol and three ('tri') fatty acids. The fatty acids can be mainly 'saturated' (in butter) or mainly 'polyunsaturated' (in margarine). There are also monounsaturated fatty acids, which occur in quantity in the triglycerides of olive oil and peanut oil. Fat is energy dense (9 kilocalories per gram), it does not mix with water, so that the food in which it is found tends to be more energy dense because of the relative lack of water. Eating fat in reasonable quantity is good for health. It must be remembered that the quality of the fat eaten as well as its quantity is important. Fat also confers texture and flavour to food, enhancing its palatability.

Types of Fats

There are several types of fats, the main ones are called triglycerides, cholesterol esters, and phospholipids. They contain fatty acids, which may be saturated, monounsaturated or polyunsaturated. Some polyunsaturated fatty acids are essential for humans because they cannot be made in our bodies. One group of essential fatty acids comes from plant sources; these fatty acids are found in considerable quantity in polyunsaturated margarine and in vegetable oils. We need about 1 to 2 per cent of our energy to come from this type of fatty acid.

Saturated fatty acids come mainly from ruminant animals, such as sheep and cattle, and from milk and dairy products of these animals. Monounsaturated fat does not increase the blood cholesterol level and may decrease it (olive oil and peanut oil provide monounsaturated fat). Polyunsaturated fat decreases it.

Fat Intake

WFP recommends that 17% of the energy of the ration is derived from fat.

For instance for a daily ration of 2,100 kcal;

- 17% = 357 kcal
- 357 kcal/ 9 kcal/g = 39.7 g of fat per day
- Maize contains 3g of fat per 100g. Beans contains 2g of fat per 100g.
- A ration made of 400g of maize (400 x 3% = 12g), 100g of beans (100 x 2% = 2g), and 25g of oil will cover the daily needs in fat (12 + 2 + 25 = 39g fat).

Carbohydrates

Carbohydrates can be divided into three main groups:

- Sugars
- Starches
- Dietary fibers.

Sugars and starches in food are sources of energy. Cellulose and some related substances are not used by our bodies as a significant source of energy. Nevertheless, these components are very important as, together with other indigestible substances, they constitute dietary fiber.

Sugars

The main sugars in food are sucrose, glucose, fructose, maltose and lactose. Sucrose is obtained from sugar cane (or sugarbeet) and is usually called 'sugar'. Besides providing energy, sugars also produce the sensation of sweetness. Each sugar contributes the same amount of energy (kilocalories) to our diet regardless of its sweetness. Different sugars are not equally sweet and the degree of sweetness of a food is often not a good indication of the amount of sugars present. For example, maltose is only half as sweet as sucrose.

Starch

Starch is the main form of carbohydrate in our food. It is found in a variety of cereals, vegetables and fruit, but mostly in cereals, flours and legumes (beans, peas). Starchy foods are usually cooked to improve digestibility and give a more desirable texture and flavour. During the ripening of fruit, starch is changed into sugars, which give sweetness to ripe fruits. In contrast to sugars, starch is often accompanied by significant amounts of other nutrients including dietary fiber. Starch has the same energy value as sugars.

Carbohydrate Intake

There is no specific dietary requirement for carbohydrate because energy can also be derived from protein, and fat. However, a diet that does not contain carbohydrate can lead to muscle breakdown and dehydration. This can be prevented by 50 to 100 grams of carbohydrate per day, but levels above this are desirable. Sources of complex carbohydrates, such as starch, are recommended as these often also provide necessary vitamins, elements (minerals) and dietary fiber.

Dietary Fibers

Dietary fibre is a term that refers to a group of food components that pass through the stomach and small intestine undigested and reach the large intestine virtually unchanged. Most other nutrients are digested and are being used in other parts of the body by this stage. During its passage through the large intestine some components of dietary fiber are broken down to varying degrees and absorbed by the body; the remaining components are excreted in the faeces. The current attention being given to the role of dietary fiber in prevention of certain diseases is largely due to the observation that patterns of disease observed in Africa and Asia were different from those in Western countries. It was suggested that the dietary fiber content was associated with this difference. Although it is not yet proven, there is evidence to suggest that dietary fiber is beneficial for a good health.

Nutritionally Improved Food Feeds

The use of modern scientific practices such as biotechnology in agriculture has made it possible to introduce a specific characteristic in a particular grain that can improve its efficiency as a livestock feed. A wide range of options has been put forward by scientists and industry specialists as potential means of improving the nutritional composition of feed grains that would address the specific needs of different livestock industries.

In assessing research priorities in the area of feed grains quality improvement, there has been a lack of information on the economics of the various research options. In recognition of that knowledge gap, the Grains Research and Development Corporation (GRDC) funded a project, "Economic assessment of improving nutritional characteristics of feed grains (DAN331A)". The project was a collaborative one under the leadership of NSW Agriculture, involving the Australian Bureau of Agricultural and Resource Economics (ABARE) and ACE Livestock Consulting Pty Ltd. That project aimed to provide for the first time a comprehensive set of information on the value of improving different characteristics of feed grains for animal nutrition, and information on who was likely to receive the benefits of the research. The objective of the analysis undertaken in this study was to assess those potential new feeds and determine the economic merit of research to develop those feeds.

A comprehensive set of options for new feed types has been evaluated, to establish the options with the highest priorities for research. In addition, to provide a benchmark for the value of the nutritional improvements, other forms of feed grains improvement were also assessed. The options analysed are classified as follows:

- Feeds involving change in protein content
- Feeds involving change in amino acid profile
- Feeds involving improvement in feed digestibility and efficiency
- Feeds involving reduction in anti-nutritional factors
- Feeds involving increase in yield
- New crop options.

The nutritional value of each of the new options was compared to the "standard" or unimproved feed grain. In some of the options, the nutritional quality of the grain can be changed without affecting its yield, and without any change in agronomic practices or the cost of production. In others, there were associated yield changes or changes in the level of inputs that would be needed to produce the nutritionally improved feed grain. In assessing the relative benefits from alternative forms of improvement of nutrition of feed grains, the cost-reducing impacts of the different options have been analysed in a linear programming model that determines the least cost feed rations for the different livestock industries. The aggregate model considers 43 feed ingredients and estimates the least cost feed rations for the 12 livestock industries simultaneously. The cost-reduction from the new feeds was identified for each livestock industry. Economic welfare analysis was then used to estimate the size and distribution of the benefits of research from the feed grains quality-improving research between the producers (including input suppliers such as grain producers) and the consumers (including processors and final consumers) of those livestock products. The analysis also identified which of the livestock industries were likely to receive the benefits from each of the new feeds.

All of the new feeds were analysed using the aggregate feed demand model, to give a comparative analysis of all the feeds. A selected subset of the new feeds was then analysed using ABARE's regional model. That analysis allowed some of the key potential new feeds to be examined in detail, while still being comparable to the full set of options. The analysis also reveals that the aggregate national analysis provides a valuable assessment of the overall value of the new feeds.

When the feeds were analysed to assess the economic benefits, a large number of the options were found to have small or very small returns that would not justify a significant research input. The analysis reveals that there are some opportunities to improve the productivity and competitiveness of Australia's livestock industries by improving the nutritional characteristics of some feed grains. The feeds that provide the largest welfare benefits are: High oil lupins and Naked oats. The potential benefits from several other feeds are also sufficient to make them worthwhile research targets in the feed grains area, including: High oil sorghum, High protein lupins, Low arabinoxylan wheat, Hull-less barley, High oil oats, Low seed coat content barley, and High seed coat digestibility barley.

However, there are a large number of technically feasible potential new feeds that are not likely to produce sufficient benefits to make them a reasonable research target. Of the 25 feeds with improved nutritional characteristics that were analysed, 10 had total welfare benefits of less than $0.3 million per year and a further 6 less than $1.2 million per year. Given the expected research costs, probabilities of success and the time lags involved in developing these feeds by plant breeding, it is unlikely that these options could be expected to provide a satisfactory rate of return on the research funds required. Research funds used for these projects could well be applied to more productive projects. Several of those leading options for nutritional improvement had negative impacts on some livestock industries, so that none were able to provide universal benefits to all the industries included in the analysis. As a result, different livestock industries would rank the potential new feeds in different ways, often markedly differently.

An alternative would be to aim for yield improvement rather than seek to improve the nutritional quality of the feeds. That direction for research funding would provide economic benefits of similar or greater size than from nutritional improvement, and the evidence from the analysis in this study is that those benefits may well be more evenly spread across the different industries. Clearly, the selection of which new feeds to develop needs to be undertaken carefully, to ensure that scarce research and development funds are used to provide the best returns. The analysis in this study enables those feeds to be identified, so that research priorities for feed grains can be developed with improved knowledge of the economic consequences.

Nutritional Assessment Process for Biotechnology

Modern crop biotechnology uses new techniques to do what farmers and plant breeders have been doing for centuries: making small changes to the genetic make-up of plants to improve growing or eating characteristics.

The new knowledge and techniques that biotechnology has brought to plant breeding have significantly improved the precision and the speed of the breeding process, thereby significantly improving the efficiency with which scientists can develop new crop varieties for Asia. Genetically modified foods are the most studied food products ever produced, but it is sometimes difficult for consumers to understand the exhaustive checking and testing processes that are applied to biotechnology-derived foods. Improving public knowledge of the safety assessment of biotechnology- derived foods is now thought to be an even greater priority and challenge than further development of the science and technology itself.

What people in Asia think about biotechnology AFIC's own consumer research found that ordinary people feel poorly informed about this topic area. They would like to know more, but in language and terms that are understandable and acceptable to scientists and non-scientists alike. Indeed consumers have a right to know that their food is safe and nutritious. However, in order for consumer to feel confident that foods derived from modern biotechnology methods are safe to eat, they need some basic understanding of the safety tests and precautions that are applied the new varieties of crops and produce that are becoming available. New technologies, new language and knowledge As personal computers and the worldwide web emerged, a whole new family of words and phrases also emerged, which consumers have quickly adopted and become quite comfortable with. Similarly, modern biotechnology brings with it new language and new knowledge needs. One example of this is 'substantial equivalence'. This is a term unfamiliar to most non-specialists, but which is fundamental to the principle of safety assessment for genetically modified foods. The Food and Agriculture Organization (FAO) and the World Health Organization (WHO) of the United Nations advocate the concept of 'substantial equivalence' as the most practical approach to address the safety evaluation of foods or food components derived by modern biotechnology.

In many Asian countries, the most readily understandable terms for foods derived from modern biotechnology breeding methods are food biotechnology or genetically modified foods. Abbreviations such as 'GM food' or GMOs are perceived as jargon. Such terminology may lead to confusion, miscommunication and even misinterpretation of the topic and related issues.

What is Substantial equivalence ? Substantial equivalence is based on the principle that, 'if a new food or food component is found to be substantially equivalent to an existing food or food component, it can be treated in the same manner with respect to safety'. For a foodstuff to be assessed as substantially equivalent to currently available products, the products in question is subjected to multiple tests and checks. These include molecular characterisation of the genetic modification, agronomic characterisation, nutritional assessment, toxicological assessment and safety assessment. For example, typical questions that have to be addressed are:

- Does the genetically modified food have a traditional counterpart that has a history of safe use?
- Has the concentration of any naturally occurring toxins or allergens in the food changed?
- Have the levels of key nutrients changed?
- Do new substances in the genetically modified food have a history of safe use?
- Has the food's digestibility been affected?
- Has the food been produced using accepted, established procedures?
- The overall goal of these tests is to determine whether the plant is substantially equivalent (in terms of chemical and nutritional composition and characteristics) to food derived from a conventional source that has a history of safe use.

A substantial equivalence evaluation focuses on the product rather than the process used to develop the product. If the new product is substantially equivalent to the conventional food or feed, then the product derived through biotechnology is considered to be as safe as the conventional counterpart. If the food produced using biotechnology contains a new trait, which changes the levels of nutrients or anti-nutrients, such as a higher level of a vitamin or a lower level of an allergen, the assessment focuses on demonstrating the safety of the new trait. Researchers must prepare comprehensive data to support the safety and holesomeness of new crop varieties developed through biotechnology. This process requires years of laboratory and field testing before a product can be brought to the market. This article is based on extracts from Food Biotechnology: a Communications Guide to Enhance Understanding. The Guide has been produced by the Asian Food Information Centre (AFIC) and the International Service for the Acquisition of Agri-biotech Applications (ISAAA). It is intended to provide anyone who needs to write or talk about biotechnology or who simply wants to understand the science and the issues related to this important topic area, with the necessary information resources.

The kit is designed to provide the most scientifically sound and up-to-date information about biotechnology products and processes, in language that both scientists and non-scientists can understand and agree upon. The guide is available in English and other languages. Countries in Asia have a long history of producing foods using biotechnology including soy sauce, tempeh and natto (fermented soybeans), belacan (fermented shrimp paste), cincaluk (fermented shrimps), budu and ngoc nam (fermented fish sauce), tapai (fermented milk), toddy (fermented young flowers of palm) and sake. Foods such as pickles, vinegar, bread, yoghurt and cheese are also the products of biotechnology.

Modernadvances in biotechnology are not a panacea for all the challenges that Asia currently faces in providing its growing population with a varied, safe and high guality food supply. Nevertheless, biotechnology could contribute to this goal, through the development of crops that can give improved yields, require less pesticides, result in less enviromental degradation, better nutritional profiles, and better keeping and eating qualities to name but name but a few of the potential benefits.

Some of the tests applied to improved crop varieties, to determine whether the plant is substantially equivalent (in terms of chemical and nutritional composition and characteristics) to food derived from a conventional source that has a history of safe use.

Safety assessment- Molecular characterisation - for new plant varieties produced through modern biotechnology, the source of the gene introduced into the plant is first identified. The transformation system used to insert the gene is defined as well as the number of copies of inserted genes and the integrity and stability of the genetic insert.

Agronomic traits - Usually the starting points for evaluating substantial equivalence. For example, in the case of potatoes, the traits commonly examined are yield, tuber size and distribution, dry matter content and disease resistance.

Nutritional assessment - Involves key nutrients including fats, proteins, carbohydrates and essential vitamins and minerals.

Toxicology assessment - Toxicants and anti-nutrients are compounds known to be naturally present in some crops that could have an impact on health if their levels increased. For example, solanine glycoalkaloids in potatoes or trypsin inhibitors in soybeans. The levels of antinutrients in crops produced through biotechnology are compared to conventionally produced varieties grown under comparable environmental and agronomic conditions.

Allergenicity - Genes from common allergenic foods are not used. Allergic responses to foods are almost always due to protein molecules in the food. Tests include examination of molecular structure, stability of protein in stomach and intestinal fluids and measurement of the amount of any new protein in the food.

The Concept of Substantial Equivalence, its Historical

Development and Current Use

For many years the practical difficulties of obtaining meaningful information on the safety of whole foods from conventional toxicology studies have been well recognized. This became particularly apparent from the vast number of animal feeding studies conducted to assess the safety of irradiated foods.

Animal studies are a major element in the safety assessment of many compounds such as pesticides, pharmaceuticals, industrial chemicals and food additives. In most cases however, the test substance is well characterised, of known purity, of no nutritional value and human exposure is generally low. It is therefore relatively straightforward to feed such compounds to animals at a range of doses, some several orders of magnitude greater than the expected human exposure levels, in order to identify any potential adverse effects of importance to humans. In this way it is possible, in most cases, to determine levels of exposure at which adverse effects are not present, and so set safe upper limits by the application of appropriate safety factors. By contrast, foods are complex mixtures of compounds characterised by wide variation in composition and nutritional value. Due to their bulk and effect on satiety they can usually only be fed to animals at low multiples of the amounts that might be present in the human diet. In addition, a key factor to consider in conducting animal studies on foods is the nutritional value and balance of the diets used, to try to avoid the induction of adverse effects which are not related directly to the material itself. Picking up any potential adverse effects and relating these conclusively to an individual characteristic of the food can therefore be extremely difficult. Another consideration in deciding the need for animal studies is whether it is appropriate to subject experimental animals to such a study if it is unlikely to give rise to meaningful information. In practice very few foods consumed today have been subject to any toxicological studies yet are generally accepted as being safe to eat. In developing a methodology for the safety assessment of new foods it was essential to establish a benchmark definition of safe food.

Development of a Safety Assessment Framework for GM Foods

Recognising that the development of GM foods was progressing rapidly the FAO and WHO convened an expert consultation in 1990 on the 'Assessment of Biotechnology in Food Production and Processing as Related to Food Safety'. The consultation recognised the limitations of traditional toxicological test methods when applied to whole foods and recommended that a more structured approach to safety assessment should be developed. The 1990 consultation identified the comparative principle whereby the food being assessed is compared with one that has an accepted level of safety, as being of considerable importance.

In 1993 the OECD published a report on the safety evaluation of foods derived by modern biotechnology. This report which was based on a number of intergovernmental consultations included a definition of safe food.

"Food is considered safe if there is reasonable certainty that no harm will result from its consumption under anticipated conditions. Historically, food prepared and used in traditional ways is considered safe on the basis of long term experience, even though it may naturally contain harmful substances. In principle, food is presumed to be safe unless a significant hazard has been identified."

This is not to say that many foods that are already widely consumed would not show adverse effects in animal studies if they could be fed at high enough doses. Equally given the many adverse effects that can be observed with existing foods it would be unreasonable to require a demonstration of absolute safety for novel foods.

The 1993 OECD report also expanded upon the comparative principle identified by the 1990 FAO/WHO consultation and formulated the concept of substantial equivalence.

The WHO and FAO refined the concept further at an expert consultation meeting held in Rome in 1996.

What is Substantial Equivalence

Substantial equivalence is not a substitute for a safety assessment, but a part of the assessment process. As such, it provides a useful framework for regulatory scientists. Underlying the concept is the requirement that any safety assessment should show that a genetically modified variety is as safe as its traditional counterparts, through a consideration of both intended and unintended effects. This involves consideration of a wide range of information, including agronomic properties, phenotypic changes and compositional data on critical nutrients and toxicants.

In the report of the 1996 expert consultation substantial equivalence was identified as being 'established by a demonstration that the characteristics assessed for the genetically modified organism, or the specific food product derived therefrom, are equivalent to the same characteristics of the conventional comparator. The levels and variation for characteristics in the genetically modified organism must be within the natural range of variation for those characteristics considered in the comparator and be based upon an appropriate analysis of data.'

Critical nutrients and toxicants are components of a particular crop known to be relevant to human health, as determined through our knowledge of the unmodified crop and its related species. Comparative assessment of these components and their potential for change as a result of genetic modification, together with a wide range of other information on agronomic, phenotypic and other properties, permits an assessment of the likelihood of unintended effects in a modified crop. For example, where the level of expression of a particular gene is altered, this is likely to be reflected in other changes in either the crop's composition or appearance.

In the application of substantial equivalence, these key components are Considered on the basis of the long history of safe use of the traditional counterpart and any differences are identified. The defined differences are then the subject of safety assessment, which can include nutritional, toxicological and immunological testing as appropriate.

The concept of substantial equivalence has been used extensively as a tool in assessing the safety of GM foods. In comparing a GM food with a conventional counterpart, consideration is given to both intentional and unintentional effects. A wide range of information is used in these comparisons ranging from agronomic data such as crop height, yield, flowering pattern, disease resistance etc. through to compositional data on key nutrients and toxicants. In this context, key nutrients are those food components which may have a major impact on the total diet and include fats, proteins and carbohydrates as well as minerals and vitamins. The comparison can result in one of three conclusions:

- The GMO or food product obtained from it is substantially equivalent to a conventional counterpart.
- The GMO or food product obtained from it is substantially equivalent to a conventional counterpart except for a few clearly defined differences.
- The GMO or food product obtained from it is not substantially equivalent to a conventional counterpart - either because the differences cannot be defined or because there is no existing counterpart to compare it with.

Where a food can be demonstrated to be substantially equivalent, it is considered to be as safe as its counterpart and no further safety assessment is required. Where there are clearly defined differences between the GM food and its conventional counterpart, the safety implications of the differences need to be fully assessed. Where a food is not substantially equivalent, it does not mean that the food is unsafe. However, there would be a need for extensive data to be provided to demonstrate its safety.

How Substantial Equivalence is Currently Used in Practice?

The concept of substantial equivalence has been integrated into safety assessment procedures used by regulatory authorities worldwide. In many countries e.g. the EU, Australia, Canada and the USA the safety assessment process has been formalised in a series of decision trees which guide regulators and potential applicants through the various stages of the comparative process.

At present each regulatory authority could, in theory, determine a unique set of components they wish to see analysed. In practice there is already broad agreement on key components, although work is progressing to develop international consensus on a core set of components on a crop by crop basis. Although the concept has been interpreted in slightly different ways by various regulatory authorities, the overall approach to the safety assessment is very similar in all countries. The area where there is perhaps the greatest scope for divergence is which of the three categories identified in paragraph 14 that a particular novel food is assigned to. This is largely down to differing interpretations as to what constitutes a difference from a conventional counterpart.

Whilst the foods derived from biotechnology which are on sale have all been assessed for safety using the substantial equivalence concept and the results endorsed by the respective Governments, there have been some published criticisms of substantial equivalence. Many of these criticisms are based on a misunderstanding of the concept. Nevertheless such criticisms provide a useful stimulus to ensure that safety assessment procedures are kept at the forefront of scientific knowledge.

The concept of substantial equivalence has been used by regulatory authorities worldwide for approximately ten years. In this time some 40 products have been assessed, the complexity of the genetic modifications has increased. For some products, particularly commodity crops such as soya, many millions of tonnes have been consumed with no evidence of any adverse health effects. The concept has stood the test of time, although it is important that it is kept under review to ensure it remains the most appropriate mechanism for assessing the nutritional and food safety implications of foods derived from biotechnology.

Chapter 3

Food Processing

Food is any substance or materials that is consumed to provide nutritional support for the body or for pleasure. It is usually of plant or animal origin, and contains essential nutrients, such as carbohydrates, fats, proteins, vitamins, or minerals. It is ingested and assimilated by an organism to produce energy, stimulate growth, and maintain life.

Historically, people obtained food from hunting and gathering, farming, ranching, and fishing, known as agriculture. Today, most of the food energy consumed by the world population is supplied by the food industry operated by multinational corporations using intensive farming and industrial agriculture methods.

Food safety and food security are monitored by agencies such as the International Association for Food Protection, World Resources Institute, World Food Programme, Food and Agriculture Organization, and International Food Information Council. They address issues such as sustainability, biological diversity, climate change, nutritional economics, population growth, water supply, and access to food.

The right to food is a human right derived from the International Covenant on Economic, Social and Cultural Rights (ICESCR), recognizing the "right to an adequate standard of living, including adequate food", as well as the "fundamental right to be free from hunger".

Food Sources

Almost all foods are of plant or animal origin. Cereal grain is a staple food that provides more food energy worldwide than any other type of crop. Maize, wheat, and rice together account for 87% of all grain production worldwide.

Other foods not from animal or plant sources include various edible fungi, especially mushrooms. Fungi and ambient bacteria are used in the preparation of fermented and pickled foods such as leavened bread, alcoholic drinks, cheese, pickles, kombucha, and yogurt. Another example is blue-green algae such as Spirulina. Inorganic substances such as baking soda and cream of tartar are also used to chemically alter an ingredient.

Plants

Many plants or plant parts are eaten as food. There are around 2,000 plant species which are cultivated for food, and many have several distinct cultivars.

Seeds of plants are a good source of food for animals, including humans, because they contain the nutrients necessary for the plant's initial growth, including many healthy fats, such as Omega fats. In fact, the majority of food consumed by human beings are seed-based foods. Edible seeds include cereals (such as maize, wheat, and rice), legumes (such as beans, peas, and lentils), and nuts. Oilseeds are often pressed to produce rich oils, such as sunflower, flaxseed, rapeseed (including canola oil), and sesame. Seeds are typically high in unsaturated fats and, in moderation, are considered a health food, although not all seeds are edible. Large seeds, such as those from a lemon, pose a choking hazard, whereas seeds from apples and cherries contain the poison cyanide.

Fruits are the ripened ovaries of plants, including the seeds within. Many plants have evolved fruits that are attractive as a food source to animals, so that animals will eat the fruits and excrete the seeds some distance away. Fruits, therefore, make up a significant part of the diets of most cultures. Some botanical fruits, such as tomatoes, pumpkins, and eggplants, are eaten as vegetables.

Vegetables are a second type of plant matter that is commonly eaten as food. These include root vegetables (such as potatoes and carrots), leaf vegetables (such as spinach and lettuce), stem vegetables (such as bamboo shoots and asparagus), and inflorescence vegetables (such as globe artichokes and broccoli). Many herbs and spices are highly flavour some vegetables.

Animals

Animals are used as food either directly or indirectly by the products they produce. Meat is an example of a direct product taken from an animal, which comes from either muscle systems or from organs. Food products produced by animals include milk produced by mammary glands, which in many cultures is drunk or processed into dairy products such as cheese or butter. In addition, birds and other animals lay eggs, which are often eaten, and bees produce honey, a reduced nectar from flowers, which is a popular sweetener in many cultures. Some cultures consume blood, sometimes in the form of blood sausage, as a thickener for sauces, or in a cured, salted form for times of food scarcity, and others use blood in stews such as civet.

Some cultures and people do not consume meat or animal food products for cultural, dietary, health, ethical, or ideological reasons. Vegetarians do not consume meat. Vegans do not consume any foods that are or contain ingredients from an animal source.

Food Chain

Food chains and food webs are representations of the predator-prey relationships between species within an ecosystem or habitat.

Many chain and web models can be applicable depending on habitat or environmental factors. Every known food chain has a base made of autotrophs, organisms able to manufacture their own food (e.g. plants, chemotrophs).

Organisms Represented in Food Chains

In nearly all food chains, solar energy is input into the system as light and heat, utilized by autotrophs (i.e., producers) in a process called photosynthesis. Carbon dioxide is reduced (gains electrons) by being combined with water (a source of hydrogen atoms), producing glucose. Water splitting produces hydrogen, but is a nonspontaneous (endergonic) reaction requiring energy from the sun. Carbon dioxide and water, both stable, oxidized compounds, are low in energy, but glucose, a high-energy compound and good electron donor, is capable of storing the solar energy. This energy is expended for cellular processes, growth, and development. The plant sugars are polymerized for storage as long-chain carbohydrates, including other sugars, starch, and cellulose.

Glucose is also used to make fats and proteins. Proteins can be made using nitrates, sulfates, and phosphates in the soil. When autotrophs are eaten by heterotrophs, i.e., consumers such as animals, the carbohydrates, fats, and proteins contained in them become energy sources for the heterotrophs.

Chemoautotrophy

An important exception is lithotrophy, the utilization of inorganic compounds, especially minerals such as sulphur or iron, for energy. In some lithotrophs, minerals are used simply to power processes for making organic compounds from inorganic carbon sources. In a few food chains, e.g., near hydrothermal vents in the deep sea, autotrophs are able to produce organic compounds without sunlight, through a process similar to photosynthesis called chemosynthesis, using a carbon source such as carbon dioxide and a chemical energy sources such as hydrogen sulfide, H_2S, or molecular hydrogen, H_2.

Unlike water, the hydrogen compounds used in chemosynthesis are high in energy. Other lithotrophs are able to directly utilize inorganic substances, e.g., iron, hydrogen sulfide, elemental sulphur, or thiosulfate, for some or all of their energy needs.

Involvement in the Carbon Cycle

Carbon dioxide is recycled in the carbon cycle as carbohydrates, fats, and proteins are oxidized (burned) to produce carbon dioxide and water. Oxygen released by photosynthesis is utilized in respiration as an electron acceptor to release chemical energy stored in organic compounds. Dead organisms are consumed by detritivores, scavengers, and decomposers, including fungi and insects, thus returning nutrients to the soil.

Food Web

Food chains are overly simplistic as representatives of the relationships of living organisms in nature. Most consumers feed on multiple species and in turn, are fed upon by multiple other species.

For a snake, the prey might be a mouse, a lizard, or a frog, and the predator might be a bird of prey or a badger. The relations of detritivores and parasites are seldom adequately characterized in such chains as well.

A food web is a series of related food chains displaying the movement of energy and matter through an ecosystem. The food web is divided into two broad categories: the grazing web, beginning with autotrophs, and the detrital web, beginning with organic debris. There are many food chains contained in these food webs.

In a grazing web, energy and nutrients move from plants to the herbivores consuming them to the carnivores or omnivores preying upon the herbivores. In a detrital web, plant and animal matter is broken down by decomposers, e.g., bacteria and fungi, and moves to detritivores and then carnivores.

There are often relationships between the detrital web and the grazing web. Mushrooms produced by decomposers in the detrital web become a food source for deer, squirrels, and mice in the grazing web. Earthworms eaten by robins are detritivores consuming decaying leaves.

Flow of Food Chains

Food energy flows from one organism to the next and to the next and so on, with some energy being lost at each level. Organisms in a food chain are grouped into trophic levels, based on how many links they are removed from the primary producers. In trophic levels there may be one species or a group of species with the same predators and prey.

Autotrophs such as plants or phytoplankton are in the first trophic level; they are at the base of the food chain. Herbivores (primary consumers) are in the second trophic level. Carnivores (secondary consumers) are in the third. Omnivores are found in the second and third levels. Predators preying upon other predators are tertiary consumers or secondary carnivores, and they are found in the fourth trophic level.

Food chain length is another way of describing food webs as a measure of the number of species encountered as energy or nutrients move from the plants to top predators. There are different ways of calculating food chain length depending on what parameters of the food web dynamic are being considered: connectance, energy, or interaction. In a simple predator-prey example, a deer is one step removed from the plants it eats (chain length = 1) and a wolf that eats the deer is two steps removed (chain length = 2). The relative amount or strength of influence that these parameters have on the food web address questions about:

- the identity or existence of a few dominant species (called strong interactors or keystone species)
- the total number of species and food-chain length (including many weak interactors) and
- how community structure, function and stability is determined.

Trophic Level

The trophic level of an organism is the position it occupies on the food web. The word trophic derives from the Greek τροφη (trophe) referring to food or feeding. A food chain represents a succession of organisms that eat another organism and are, in turn, eaten themselves. The number of steps an organism is from the start of the chain is a measure of its trophic level. Food chains start at trophic level 1 with primary producers such as plants, move to herbivores at level 2, predators at level 3 and typically finish with carnivores or apex predators at level 4 or 5. The path along the chain can form a one-way flow, or a food "web." Ecological communities with higher biodiversity form more complex trophic paths.

The plants in this image, and the algae and phytoplankton in the lake, are primary producers. They take nutrients from the soil or the water, and manufacture their own food by photosynthesis, using energy from the sun. The three basic ways organisms get food are as producers, consumers and decomposers.

- Producers (autotrophs) are typically plants or algae. Plants and algae do not usually eat other organisms, but pull nutrients from the soil or the ocean and manufacture their own food using photosynthesis. For this reason, they are called primary producers. In this way, it is energy from the sun that usually powers the base of the food chain. An exception occurs in deep-sea hydrothermal ecosystems, where there is no sunlight. Here primary producers manufacture food through a process called chemosynthesis.
- Consumers (heterotrophs) are animals which cannot manufacture their own food and need to consume other organisms. Animal that eat primary producers (like plants) are called herbivores. Animals that eat other animals are called carnivores, and animals that eat both plant and other animals are called omnivores.
- Decomposers (detritivores) break down dead plant and animal material and wastes and release it again as energy and nutrients into the ecosystem for recycling. Decomposers, such as bacteria and fungi (mushrooms), feed on waste and dead matter, converting it into inorganic chemicals that can be recycled as mineral nutrients for plants to use again.

Trophic levels can be represented by numbers, starting at level 1 with plants. Further trophic levels are numbered subsequently according to how far the organism is along the food chain.

- Level 1: Plants and algae make their own food and are called primary producers.
- Level 2: Herbivores eat plants and are called primary consumers.
- Level 3: Carnivores which eat herbivores are called secondary consumers.

- Level 4: Carnivores which eat other carnivores are called tertiary consumers.
- Level 5: Apex predators which have no predators are at the top of the food chain.

In real world ecosystems, there is more than one food chain for most organism, since most organisms eat more than one kind of food or are eaten by more than one type of predator. A diagram which sets out the intricate network of intersecting and overlapping food chains for an ecosystem is called its food web. Decomposers are often left off food webs, but if included, they mark the end of a food chain. Thus food chains start with primary producers and end with decay and decomposers. Since decomposers recycle nutrients, leaving them so they can be reused by primary producers, they are sometimes regarded as occupying their own trophic level.

Biomass Transfer Efficiency

Generally, each trophic level relates to the one below it by absorbing some of the energy it consumes, and in this way can be regarded as resting on, or supported by the next lower trophic level. Food chains can be diagrammed to illustrate the amount of energy that moves from one feeding level to the next in a food chain. This is called an energy pyramid. The energy transferred between levels can also be thought of as approximating to a transfer in biomass, so energy pyramids can also be viewed as biomass pyramids, picturing the amount of biomass that results at higher levels from biomass consumed at lower levels.

The efficiency with which energy or biomass is transferred from one trophic level to the next is called the ecological efficiency. Consumers at each level convert on average only about 10 percent of the chemical energy in their food to their own organic tissue. For this reason, food chains rarely extend for more than 5 or 6 levels. At the lowest trophic level (the bottom of the food chain), plants convert about one percent of the sunlight they receive into chemical energy. It follows from this that the total energy originally present in the incident sunlight that is finally embodied in a tertiary consumer is about 0.001 %

Fractional Trophic Levels

Food webs largely define ecosystems, and the trophic levels define the position of organisms within the webs. But these trophic levels are not always simple integers, because organisms often feed at more than one trophic level. For example, some carnivores also eat plants, and some plants are carnivores. A large carnivore may eat both smaller carnivores and herbivores; the bobcat eats rabbits, but the mountain lion eats both bobcats and rabbits. Animals can also eat each other; the bullfrog eats crayfish and crayfish eat young bullfrogs. The feeding habits of a juvenile animal, and consequently its trophic level, can change as it grows up. The fisheries scientist Daniel Pauly sets the values of trophic levels to one in plants and detritus, two in herbivores and detritivores (primary consumers), three in secondary consumers, and so on. The definition of the trophic level, TL, for any consumer species i is:

$$TL_i = 1 + \sum_j \left(TL_j.DC_{ij}\right)$$

where TL_i is the fractional trophic level of the prey j, and DC_{ij} represents the fraction of j in the diet of i. In the case of marine ecosystems, the trophic level of most fish and other marine consumers takes value between 2.0 and 5.0. The upper value, 5.0, is unusual, even for large fish, though it occurs in apex predators of marine mammals, such as polar bears and killer whales.

Mean trophic level

In fisheries, the mean trophic level for the fisheries catch across an entire area or ecosystem is calculated for year y as:

$$TL_y = \frac{\sum_i \left(TL_i.Y_{iy}\right)}{\sum_i Y_{iy}}$$

where Y_{iy} is the catch of the species or group i in year y, and TL_i is the fractional trophic level for species i as defined above.

It was once believed that fish at higher trophic levels usually have a higher economic value; resulting in overfishing at the higher trophic levels. Earlier reports found precipitous declines in mean trophic level of fisheries catch, in a process known as fishing down the food web. However, more recent work finds no relation between economic value and trophic level; and that mean trophic levels in catches, surveys and stock assessments have not in fact declined, suggesting that fishing down the food web is not a global phenomenon.

FiB Index

Since biomass transfer efficiencies are only about 10 percent, it follows that the rate of biological production is much greater at lower trophic levels than it is at higher levels. Fisheries catches, at least to begin with, will tend to increase as the trophic level declines. At this point the fisheries will target species lower in the food web. In 2000, this led Pauly and others to construct a "Fisheries in Balance" index, usually called the FiB index. The FiB index is defined, for any year y, by

$$FiBy = \log ((Y_y/(TE)^{TLy})/(Y_0/(TE)^{TL0}))$$

where Y_y is the catch at year y, TL_y is the mean trophic level of the catch at year y, Y_0 is the catch and TL_0 the mean trophic level of the catch at the start of the series being analyzed, and TE is the transfer efficiency of biomass or energy between trophic levels.

The FiB index is stable (zero) over periods of time when changes in trophic levels are matched by appropriate changes in the catch in the opposite direction. The index increases if catches increase for any reason, e.g. higher fish biomass, or geographic expansion Such decreases explain the "backward-bending" plots of trophic level versus catch originally observed by Pauly and others in 1998.

Entropic Losses in the Chain

It is the case that the biomass of each trophic level decreases from the base of the chain to the top. This is because energy is lost to the environment with each transfer as entropy increases. About eighty to ninety percent of the energy is expended for the organism's life processes or is lost as heat or waste. Only about ten to twenty percent of the organism's energy is generally passed to the next organism. The amount can be less than one percent in animals consuming less digestible plants, and it can be as high as forty percent in zooplankton consuming phytoplankton. Graphic representations of the biomass or productivity at each tropic level are called ecological pyramids or trophic pyramids. The transfer of energy from primary producers to top consumers can also be characterized by energy flow diagrams.

Ecological Efficiency

Ecological efficiency describes the efficiency with which energy is transferred from one trophic level to the next. It is determined by a combination of efficiencies relating to organismic resource acquisition and assimilation in an ecosystem.

Energy Transfer

Primary production occurs in autotrophic organisms of an ecosystem. Photoautotrophs such as vascular plants and algae convert energy from the sun into energy stored as carbon compounds. Photosynthesis is carried out in the chlorophyll of green plants. The energy converted through photosynthesis is carried through the trophic levels of an ecosystem as organisms consume members of lower trophic levels.

Primary production can be broken down into gross and net primary production. Gross primary production is a measure of the energy that a photoautotroph harvests from the sun. Take, for example, a blade of grass that takes in *x* Joules of energy from the sun. The fraction of that energy that is converted into glucose reflects the gross productivity of the blade of grass. The energy remaining after respiration is considered the net primary production. In general, gross production refers to the energy contained within an organism before respiration and net production the energy after respiration. The terms can be used to describe energy transfer in both autotrophs and heterotrophs.

Energy transfer between trophic levels is generally inefficient, such that net production at one trophic level is generally only 10% of the net production at the preceding trophic level. Due to non-predatory death, egestion, and respiration, a significant amount of energy is lost to the environment instead of being absorbed for production by consumers. The figure approximates the fraction of energy available after each stage of energy loss in a typical ecosystem, although these fractions vary greatly from ecosystem to ecosystem and from trophic level to trophic level.

Example: Assume 500 units of energy are produced by trophic level 1. One half of that is lost to non-predatory death, while the other half (250 units) is ingested by trophic level 2. One half of the amount ingested is expelled through defecation, leaving the other half (125 units) to be assimilated by the organism. Finally one half of the remaining energy is lost through respiration while the rest (63 units) is used for growth and reproduction. This energy expended for growth and reproduction constitutes to the net production of trophic level 1, which is equal to 500 * 1/2 * 1/2 * 1/2 = 63 units.

Quantifying Ecological Efficiency

Ecological efficiency is a combination of several related efficiences that describe resource utilization and the extent to which resources are converted into biomass.

- Exploitation efficiency is the amount of food ingested divided by the amount of prey production (I/P_n)
- Assimilation efficiency is the amount of assimilation divided by the amount of food ingestion (A/I)
- Net Production efficiency is the amount of consumer production divided by the amount of assimilation (P_{n+1}/A)
- Gross Production efficiency is the assimilation efficiency multiplied by the net production efficiency, which is equivalent

to the amount of consumer production divided by amount of ingestion (P_{n+1}/I)

- Ecological efficiency is the exploitation efficiency multiplied by the assimilation efficiency multiplied by the net production efficiency, which is equivalent to the amount of consumer production divided by the amount of prey production (P_{n+1}/P_n).

Theoretically, it is easy to calculate ecological efficiency using the mathematical relationships above. It is often difficult, however, to obtain accurate measurements of the values involved in the calculation. Assessing ingestion, for example, requires knowledge of the gross amount of food consumed in an ecosystem as well as its caloric content. Such a measurement is rarely better than an educated estimate, particularly with relation to ecosystems that are largely inaccessible to ecologists and tools of measurement. The ecological efficiency of an ecosystem is as a result often no better than an approximation. On the other hand, an approximation may be enough for most ecosystems, where it is important not to get an exact measure of efficiency, but rather a general idea of how energy is moving through its trophic levels.

Applications

In agricultural environments, maximizing energy transfer from producer (food) to consumer (livestock) can yield economic benefits. A sub-field of agricultural science has emerged that explores methods of monitoring and improving ecological and related efficiencies.

In comparing the net efficiency of energy utilization by cattle, breeds historically kept for beef production, such as the Hereford, outperformed those kept for dairy production, such as the Holstein, in converting energy from feed into stored energy as tissue. This is a result of the beef cattle storing more body fat than the dairy cattle, as energy storage as protein was at the same level for both breeds. This implies that cultivation of cattle for slaughter is a more efficient use of feed than is cultivation for milk production.

While it is possible to improve the efficiency of energy use by livestock, it is vital to the world food question to also consider the differences between animal husbandry and plant agriculture. Caloric concentration in fat tissues are higher than in plant tissues, causing high-fat organisms to be most energetically-concentrated; however, the energy required to cultivate feed for livestock is only partially converted into fat cells. The rest of the energy input into cultivating feed is respirated or egested by the livestock and unable to be used by humans. Out of a total of 96.8 * 10^15 BTU of energy used in the US in 1999,

10.5% was used in food production, with the percentage accounting for food from both producer and primary consumer trophic levels. In comparing the cultivation of animals versus plants, there is a clear difference in magnitude of energy efficiency. Edible kilocalories produced from kilocalories of energy required for cultivation are: 18.1% for chicken, 6.7% for grass-fed beef, 5.7% for farmed salmon, and 0.9% for shrimp. In contrast, potatoes yield 123%, corn produce 250%, and soy results in 415% of input calories converted to calories able to be utilized by humans. This disparity in efficiency reflects the reduction in production from moving up trophic levels. Thus, it is more energetically efficient to form a diet from lower trophic levels.

Pyramids

In a pyramid of numbers, the number of consumers at each level decreases significantly, so that a single top consumer, (e.g., a polar bear or a human), will be supported by a million separate producers. There is usually a maximum of four or five links in a food chain, although food chains in aquatic ecosystems are frequently longer than those on land. Eventually, all the energy in a food chain is lost as heat.

Some producers, especially phytoplankton, are able to reproduce quickly enough to support a larger biomass of grazers. This is called an inverted pyramid, caused by a longer lifespan and slower growth rate in the consumers than in the organisms being consumed, with phytoplankton living just a few days, compared to several weeks for the zooplankton eating the phytoplankton and years for fish eating the zooplankton. A pyramid of energy, reflecting the energy or kilojoules in each level, is representative of the true relationships of the phytoplankton, zooplankton, and fish, showing phytoplankton as the largest section, then zooplankton as a smaller section, and fish as the smallest section.

History of Food Webs

Food webs serve as a framework to help ecologists organize the complex network of interactions among species observed in nature. The earliest description of a food chain was given by the medieval Afro-Arab biologist Al-Jahiz (781-868). Perhaps the earliest graphical depiction of a food web was by Lorenzo Camerano in 1880, followed independently by those of Pierce and colleagues in 1912 and Victor Shelford in 1913. Two food webs about herring were produced by Victor Summerhayes and Charles Elton and Alister Hardy in 1923 and 1924. After Charles Elton's use of food webs in his 1927 synthesis, they became a central concept in the field of ecology. The utilization of the common currency of energy flow along links in a flow was emphasized in Raymond Lindeman's work, initiating the extensive analysis of energy and material flows that are a core activity of ecosystem ecology.

Interest in food webs increased after Robert Paine's experimental and descriptive study of intertidal shores suggesting that food web complexity was key to maintaining species diversity and ecological stability. Many theoretical ecologists, including Sir Robert May and Stuart Pimm, were prompted by this discovery and others to examine the mathematical properties of food webs. According to their analyses, complex food webs should be highly unstable. The apparent paradox between the complexity of food webs observed in nature and the mathematical fragility of food web models is currently an area of intensive study and debate. The paradox may be due partially to conceptual differences between persistence of a food web and equilibrial stability of a food web. Current research points to important roles of non-random structure in the connections within the food web that develop as food webs assemble over long periods of time, of patterns in the strengths of interactions among species within the food web, of variable strengths of species interactions as species abundances change, and of spatial variation in the environment creating food webs of different structures that are connected by movement of individuals and materials, in the creation and persistence of complex food webs.

Animal Source Foods

Animal source foods (ASF) include any food item that comes from an animal source such as meat, milk, fish, eggs, cheese and yogurt. Many individuals do not consume ASF or consume little ASF by either personal choice or necessity as ASF may not be accessible or available to these people.

Nutrition of Animal Source Foods

Aside from performed vitamin A, vitamin B_{12} and vitamin D, all vitamins found in animal source foods may also be found in plant-derived foods. Examples are tofu to replace meat (both contain protein in sufficient amounts), and certain seaweeds and vegetables as respectively kombu and kale to replace dairy foods as milk (both contain calcium in sufficient amounts). Some plant-derived foods are nutrient-denser than their animal-derived counterparts (e.g. tofu).

Most humans eat an omnivorous diet (comprising animal source foods and plant source foods) though some civilisations have eaten only animal foods. Although a healthy diet containing all essential macro and micronutrients may be possible by only consuming a plant based diet (with vitamin B_{12} obtained from supplements if no animal sourced foods are consumed), some populations are unable to consume an adequate quantity or variety of these plant based items to obtain appropriate amounts of nutrients, particularly those that are found in high concentrations in ASF. Frequently, the most vulnerable populations to these micronutrient deficiencies are pregnant women, infants, and children in developing countries. In the 1980s the Nutrition Collaborative Research Support Program (NCRSP) found that six micronutrients were low in the mostly vegetarian diets of children in malnourished areas of Egypt, Mexico, and Kenya. These six micronutrients are vitamin A, vitamin B_{12}, riboflavin, calcium, iron and zinc. ASF are the only food source of Vitamin B_{12}. ASF also provide high biological value protein, energy, fat compared with plant food sources.

Health Impacts of Micronutrient Deficiency

All six micronutrients richly found in ASF, vitamin A, vitamin B_{12}, riboflavin, calcium, iron and zinc play a critical role in the growth and development of children. Inadequate stores of these micronutrients, either resulting from inadequate intake or poor absorption, is associated with poor growth, anemias (iron deficiency anemia and macrocytic anemia), rickets, night blindness, impaired cognitive functioning, neuromuscular deficits, diminished work capacity, psychiatric disorders and death. Some of these affects, such as impaired cognitive development from an iron deficiency, are irreversible.girja

Animal Source Food Supplementation

Micronutrient deficiency is associated in poor early cognitive development. Programs designed to address these micronutrient deficiencies should be targeted to infants, children, and pregnant women. To address these significant mirconutrient deficiencies, some global health researchers and practitioners developed and piloted a snack program in Kenya school children. However, some communities are vegetarians for religious or cultural reasons. Efforts must be made to develop culturally appropriate interventions to address the micronutrient deficiencies in these populations, such as through food fortification.

Animal Source Food Production

According to a 2006 United Nations initiative, the livestock industry sector emerges as one of the top two or three most significant contributors to the most serious environmental problems, at every scale from local to global." As such, using plant-derived foods is always better in the interest of the environment. Despite this, elevation of certain animals are (somewhat) more environmental than others. According to the Farralones Institute, elevation of rabbits, and chicken (on a well-considered approach) can still be done quite environmental. As such, meat and other produce as eggs may still be produced quite environmental (if this is done on an industrial, high-efficiency manner). In addition, the elevation of goats (and their produce as goat milk and meat) can too be done quite environmentally and has been favored by certain environmental activists as Mohandas Gandhi.

Production

Traditionally, food was obtained through agriculture. With increasing concern in agri-business over multinational corporations owning the world food supply through patents on genetically modified food, there has been a growing trend toward sustainable agricultural practices. This approach, partly fueled by consumer demand, encourages biodiversity, local self-reliance and organic farming methods. Major influences on food production include international organizations (e.g. the World Trade Organization and Common Agricultural Policy), national government policy (or law), and war.

In popular culture, the mass production of food, specifically meats such as chicken and beef, has come under fire from various documentaries, most recently Food, Inc, documenting the mass slaughter and poor treatment of animals, often for easier revenues from large corporations. Along with a current trend towards environmentalism, people in Western culture have had an increasing trend towards the use of herbal supplements, foods for a specific group of person (such as dieters, women, or athletes), functional foods (fortified foods, such as omega-3 eggs), and a more ethnically diverse diet.

Cuisine Preparation

Many cultures have a recognizable cuisine, a specific set of cooking traditions using various spices or a combination of flavours unique to that culture, which evolves over time. Other differences include preferences (hot or cold, spicy, etc.) and practices, the study of which is known as gastronomy. Many cultures have diversified their foods by means of preparation, cooking methods, and manufacturing. This also includes a complex food trade which helps the cultures to economically survive by way of food, not just by consumption. Some popular types of ethnic foods include Italian, French, Japanese, Chinese, American, Cajun, Thai, and Indian cuisine.

Various cultures throughout the world study the dietary analysis of food habits. While evolutionarily speaking, as opposed to culturally, humans are omnivores, religion and social constructs such as morality, activism, or environmentalism will often affect which foods they will consume. Food is eaten and typically enjoyed through the sense of taste, the perception of flavour from eating and drinking. Certain tastes are more enjoyable than others, for evolutionary purposes.

Taste Perception

Animals, specifically humans, have five different types of tastes: sweet, sour, salty, bitter, and umami. As animals have evolved, the tastes that provide the most energy (sugar and fats) are the most pleasant to eat while others, such as bitter, are not enjoyable. Water, while important for survival, has no taste. Fats, on the other hand, especially saturated fats, are thicker and rich and are thus enjoyable to eat.

Sweet

Generally regarded as the most pleasant taste, sweetness is almost always caused by a type of simple sugar such as glucose or fructose, or disaccharides such as sucrose, a molecule combining glucose and fructose. Complex carbohydrates are long chains and thus do not have the sweet taste. Artificial sweeteners such as sucralose are used to mimic the sugar molecule, creating the sensation of sweet, without the calories. Other types of sugar include raw sugar, which is known for its amber color, as it is unprocessed. As sugar is vital for energy and survival, the taste of sugar is pleasant.

The stevia plant contains a compound known as steviol which, when extracted, has 300 times the sweetness of sugar while having minimal impact on blood sugar.

Sour

Sourness is caused by the taste of acids, such as vinegar or ethanol in alcoholic beverages. Sour foods include citrus, specifically lemons, limes, and to a lesser degree oranges. Sour is evolutionarily significant as it is a sign for a food that may have gone rancid due to bacteria. Many foods, however, are slightly acidic, and help stimulate the taste buds and enhance flavour.

Salty

Saltiness is the taste of alkali metal ions such as sodium and potassium. It is found in almost every food in low to moderate proportions to enhance flavour, although to eat pure salt is regarded as highly unpleasant. There are many different types of salt, with each having a different degree of saltiness, including sea salt, fleur de sel, kosher salt, mined salt, and grey salt. Other than enhancing flavour, its significance is that the body needs and maintains a delicate electrolyte balance, which is the kidney's function.

Salt may be iodized, meaning iodine has been added to it, a necessary nutrient that promotes thyroid function. Some canned foods, notably soups or packaged broths, tend to be high in salt as a means of preserving the food longer. Historically speaking, salt has been used as a meat preservative as salt promotes water excretion, thus working as a preservative. Similarly, dried foods also promote food safety.

Bitter

Bitterness is a highly unpleasant sensation characterized by having a sharp, pungent taste. Dark, unsweetened chocolate, caffeine, lemon rind, and some types of fruit are known to be bitter.

Umami

Umami, the Japanese word for delicious, is the least known in Western popular culture, but has a long tradition in Asian cuisine. Umami is the taste of glutamates, especially monosodium glutamate or MSG. It is characterized as savory, meaty, and rich in flavour. Salmon and mushrooms are foods high in umami. Meat and other animal by-products are described as having this taste.

Presentation

Food Presentation is the art of modifying, processing, arranging, or decorating food to enhance its aesthetic appeal.

The visual presentation of foods is often considered by chefs at many different stages of food preparation, from the manner of tying or sewing meats, to the type of cut used in chopping and slicing meats or vegetables, to the style of mold used in a poured dish. The food itself may be decorated as in elaborately iced cakes, topped with ornamental sometimes sculptural consumables, drizzled with sauces, sprinkled with seeds, powders, or other toppings, or it may be accompanied by edible or inedible garnishes.

The arrangement and overall styling of food upon bringing it to the plate is termed plating. Some common styles of plating include a 'classic' arrangement of the main item in the front of the plate with vegetables or starches in the back, a 'stacked' arrangement of the various items, or the main item leaning or 'shingled' upon a vegetable bed or side item. Item location on the plate is often referenced as for the face of a clock, with six o'clock the position closest to the diner.

It is known that when presented with food, the consumer "eats" first with their eyes, a universal psychological phenomenon. Food presented in a clean and appetizing way will encourage a good flavour, even if unsatisfactory.

Contrast in Texture

Texture plays a crucial role in the enjoyment of eating foods. Contrasts in textures, such as something crunchy in an otherwise smooth dish, may increase the appeal of eating it. Common examples include adding granola to yogurt, adding croutons to a salad or soup, and toasting bread to enhance its crunchiness for a smooth topping, such as jam or butter.

Contrast in Taste

Another universal phenomenon regarding food is the appeal of contrast in taste and presentation. Opposite flavours, such as sweet and saltiness, tend to go well together, such as in kettle corn and with nuts.

Food Preparation

While many foods can be eaten raw, many foods undergo some form of preparation for reasons of safety, palatability, texture, or flavour. At the simplest level this may involve washing, cutting, trimming, or adding other foods or ingredients, such as spices. It may also involve mixing, heating or cooling, pressure cooking, fermentation, or combination with other food. In a home, most food preparation takes place in a kitchen. Some preparation is done to enhance the taste or aesthetic appeal; other preparation may help to preserve the food; others may be involved in cultural identity. A meal is made up of food which is prepared to be eaten at a specific time and place.

Animal Preparation

The preparation of animal-based food usually involves slaughter, evisceration, hanging, portioning, and rendering. In developed countries, this is usually done outside the home in slaughterhouses, which are used to process animals en masse for meat production. Many countries regulate their slaughterhouses by law. For example, the United States has established the Humane Slaughter Act of 1958, which requires that an animal be stunned before killing. This act, like those in many countries, exempts slaughter in accordance to religious law, such as kosher, shechita, and dhabi%a halal. Strict interpretations of kashrut require the animal to be fully aware when its carotid artery is cut.

On the local level, a butcher may commonly break down larger animal meat into smaller manageable cuts, and pre-wrap them for commercial sale or wrap them to order in butcher paper. In addition, fish and seafood may be fabricated into smaller cuts by a fish monger. However fish butchery may be done on board a fishing vessel and quick-frozen for preservation of quality.

Cooking

Cooking is the process of preparing food with heat. Cooks select and combine ingredients using a wide range of tools and methods. In the process, the flavour, texture, appearance, and chemical properties of the ingredients can change. Cooking techniques and ingredients vary widely across the world, reflecting unique environmental, economic, and cultural traditions. Cooks themselves also vary widely in skill and training.

Preparing food with heat or fire is an activity unique to humans, and some scientists believe the advent of cooking played an important role in human evolution. Most anthropologists believe that cooking fires first developed around 250,000 years ago. The development of agriculture, commerce and transportation between civilizations in different regions offered cooks many new ingredients. New inventions and technologies, such as pottery for holding and boiling water, expanded cooking techniques. Some modern cooks apply advanced scientific techniques to food preparation.

History of Cooking

There is no clear evidence as to when cooking was invented. Primatologist Richard Wrangham stated that cooking was invented as far back as 1.8 million to 2.3 million years ago. Other researchers believe that cooking was invented as late as 40,000 or 10,000 years ago. Evidence of fire is inconclusive as wildfires started by lightning-strikes are still common in East Africa and other wild areas, and it is difficult to determine when fire was first used for cooking, as opposed to just being used for warmth or for keeping predators away. Most anthropologists contend that cooking fires began in earnest barely 250,000 years ago, when ancient hearths, earth ovens, burnt animal bones, and flint appear across Europe and the middle East. The only evidence of human use of fire more than two million years ago is burnt earth with human remains, which most anthropologists consider coincidence rather than evidence of intentional.

However, some Fire-cracked rock, such as that in Central Texas (United States) are burned rock middens, or enormous piles fire-damaged rock dated to c. 3,500 years ago. These may represent the remains of earth ovens used in cooking since they contain evidence of *Dasylirion wheeleri* bulbs and other plants. In Great Britain similar Neolithic, Bronze Age and Iron Age features exist, but are commonly called 'burnt mounds'.

Ingredients in Cooking

Most ingredients in cooking are derived from living things. Vegetables, fruits, grains and nuts come from plants, while meat, eggs, and dairy products come from animals. Mushrooms and the yeast used in baking are kinds of fungi. Cooks also utilize water and minerals such as salt. Cooks can also use wine, an alcohol-based liquid from the fermentation of juices of grapes or other fruits.

Naturally occurring ingredients contain various amounts of molecules called *proteins*, *carbohydrates* and *fats*. They also contain water and minerals. Cooking involves a manipulation of the chemical properties of these molecules.

Proteins

Edible animal material, including muscle, offal, milk, eggs and egg whites, contains substantial amounts of protein. Almost all vegetable matter (in particular legumes and seeds) also includes proteins, although generally in smaller amounts. These may also be a source of essential amino acids. When proteins are heated they become denatured and change texture. In many cases, this causes the structure of the material to become softer or more friable - meat becomes *cooked*. In some cases, proteins can form more rigid structures, such as the coagulation of albumen in egg whites. The formation of a relatively rigid but flexible matrix from egg white provides an important component of much cake cookery, and also underpins many desserts based on meringue.

Carbohydrates

Carbohydrates include the common sugar, sucrose (table sugar), a disaccharide, and such simple sugars as glucose (from the digestion of table sugar) and fructose (from fruit), and starches from sources such as cereal flour, rice, arrowroot, potato. The interaction of heat and carbohydrate is complex.

Long-chain sugars such as starch tend to break down into simpler sugars when cooked, while simple sugars can form syrups. If sugars are heated so that all water of crystallisation is driven off, then caramelization starts, with the sugar undergoing thermal decomposition with the formation of carbon, and other breakdown products producing caramel. Similarly, the heating of sugars and proteins elicits the Maillard reaction, a basic flavour-enhancing technique.

An emulsion of starch with fat or water can, when gently heated, provide thickening to the dish being cooked. In European cooking, a mixture of butter and flour called a roux is used to thicken liquids to make stews or sauces. In Asian cooking, a similar effect is obtained from a mixture of rice or corn starch and water. These techniques rely on the properties of starches to create simpler mucilaginous saccharides during cooking, which causes the familiar thickening of sauces. This thickening will break down, however, under additional heat.

Fats

Types of fat include vegetable oils and animal products such as butter and lard. Fats can reach temperatures higher than the boiling point of water, and are often used to conduct high heat to other ingredients, such as in frying or sautéing.

Water

Cooking often involves water which is frequently present as other liquids, both added in order to immerse the substances being cooked (typically water, stock or wine), and released from the foods themselves. Liquids are so important to cooking that the name of the cooking method used may be based on how the liquid is combined with the food, as in steaming, simmering, boiling, braising and blanching. Heating liquid in an open container results in rapidly increased evaporation, which concentrates the remaining flavour and ingredients - this is a critical component of both stewing and sauce making.

Vitamins and Minerals

Vitamins are materials required for normal metabolism but which the body cannot manufacture itself and which must therefore come from soil. Vitamins come from a number of sources including fresh fruit and vegetables (Vitamin C), carrots, liver (Vitamin A), cereal bran, bread, liver e (B vitamins), fish liver oil (Vitamin D) and fresh green vegetables (Vitamin K). Many minerals are also essential in small quantities including iron, calcium, magnesium and sulphur; and in very small quantities copper, zinc and selenium. The micronutrients, minerals, and vitamins in fruit and vegetables may be destroyed or eluted by cooking. Vitamin C is especially prone to oxidation during cooking and may be completely destroyed by protracted cooking.

Cooking Methods

Methods of Cooking

There are very many methods of cooking, most of which have been known since antiquity. These include baking, roasting, frying, grilling, barbecuing, smoking, boiling, steaming and braising. A more recent innovation is microwaving. Various methods use differing levels of heat and moisture and vary in cooking time. The method chosen greatly affects the end result. Some foods are more appropriate to some methods than others. Some major hot cooking techniques include:

Roasting

Roasting is a cooking method that uses dry heat, whether an open flame, oven, or other heat source. Roasting usually causes caramelization or Maillard browning of the surface of the food, which is considered a flavour enhancement. Roasting uses more indirect, diffused heat (as in an oven), and is suitable for slower cooking of meat in a larger, whole piece. Meats and most root and bulb vegetables can be roasted. Any piece of meat, especially red meat, that has been cooked in this fashion is called a roast. In addition, large uncooked cuts of meat are referred to as roasts. Roasting is a much slower method of cooking. A roast joint of meat can take one, two, even three hours to cook - the resulting meat is tender. Also, meats and vegetables prepared in this way are described as "roasted", e.g., roasted chicken or roasted squash.

Methods

For roasting, the food may be placed on a rack, in a roasting pan or, to ensure even application of heat, may be rotated on a spit or rotisserie. During oven roasting, hot air circulates around the meat, cooking all sides evenly. There are several theories for roasting meats correctly: low-temperature cooking, high-temperature cooking, and a combination of both. Each method can be suitable under appropriate circumstances.

- A low-temperature oven, 95 °C to 160 °C (200 °F to 325 °F), is best when cooking with large cuts of meat, turkey and whole chickens. This is not technically roasting temperature, but it is called slow-roasting. The benefit of slow-roasting an item is less moisture loss and a more tender product. At true roasting temperatures, 200 °C (400 °F) or more, the water inside the muscle is lost at a high rate.

- Cooking at high temperatures is beneficial if the cut is small enough—as in filet mignon or strip loin—to be finished cooking before the juices escape.
- The combination method uses high heat just at either the beginning or the end of the cooking process, with most of the cooking at a low temperature. This method produces the golden-brown texture and crust, but maintains more of the moisture than simply cooking at a high temperature, although the product will not be as moist as low-temperature cooking the whole time. Searing and then turning down to low is also beneficial when a dark crust and caramelized flavour is desired for the finished product. Note that searing in no way "locks in" moisture: moisture loss is simply a function of heat and time.

In general, in either case, the meat is removed from heat before it has finished cooking and left to sit for a few minutes, while the inside cooks further from the residual heat content, a phenomenon known as carry over cooking. The objective in any case is to retain as much moisture as possible, while providing the texture and color. During roasting, meats and vegetables are frequently basted on the surface with butter, lard, or oil to reduce the loss of moisture by evaporation. In recent times, plastic oven bags have become popular for roasts. These cut cooking times and reduce the loss of moisture during roasting, but reduce flavour development from Maillard browning. They are particularly popular for turkeys.

Until the late 19th century, roasting by dry heat in an oven was called *baking*. Roasting originally meant turning meat or a bird on a spit in front of a fire. It is one of the oldest forms of cooking known.

Traditionally recognized roasting methods consist only of baking and cooking over or near an open fire. Grilling is normally not technically a roast, since a grill (gridiron) is used (in English-speaking countries). Smoking differs from roasting because of the lower temperature and controlled smoke application.

Meat

Most meat roasts are large cuts of meat. Many roasts are tied with string prior to roasting, often using the reef knot or, in the more traditional sense, the packer's knot. Tying holds them together during roasting, keeping any stuffing inside, and keeps the roast in a round profile, which promotes even cooking. Prior to roasting in an oven, meat is generally "browned" by brief exposure to high temperature. This imparts a traditional flavour and color to the roast. Red meats such as beef, lamb, and venison, and certain game birds are often roasted to be "pink" or "rare", meaning that the center of the roast is still red. Due to food safety concerns, this practice is not recommended with pork and poultry. Although there is a growing fashion in some restaurants to serve "rose pork", temperature monitoring of the center of the roast is the only sure way to avoid foodborne disease.

Other

Roasting is a preferred method of cooking for most poultry, and certain cuts of beef, pork, or lamb. Some vegetables, such as potatoes, zucchini, pumpkin, turnips, parsnips, cauliflower, asparagus, squash, and peppers lend themselves to roasting as well. Roasted chestnuts are also a popular snack in winter.

Barbecuing – Grilling – Rotisserie – Searing

Baking

Baking is the technique of prolonged cooking of food by dry heat acting by convection, and not by radiation, normally in an oven, but also in hot ashes, or on hot stones. It is primarily used for the preparation of bread, cakes, pastries and pies, tarts, quiches, cookies and crackers. Such items are sometimes referred to as "baked goods," and are sold at a bakery. A person who prepares baked goods as a profession is called a baker. It is also used for the preparation of baked potatoes, baked apples, baked beans, some casseroles and pasta dishes such as lasagna, and various other foods, such as the pretzel.

Many commercial ovens are provided with two heating elements: one for baking, using convection and conduction to heat the food, and one for broiling or grilling, heating mainly by radiation. Meat may also be baked, but this is usually reserved for meatloaf, smaller cuts of whole meats, and whole meats that contain stuffing or coating such as breadcrumbs or buttermilk batter; larger cuts prepared without stuffing or coating are more often roasted, a similar process, using higher temperatures and shorter cooking times. Baking can sometimes be combined with grilling to produce a hybrid barbecue variant, by using both methods simultaneously or one before the other, cooking twice. Baking is connected to barbecuing because the concept of the masonry oven is similar to that of a smoke pit.

The baking process does not require any fat to be used to cook in an oven. Some makers of snacks such as potato chips or crisps have produced baked versions of their snack items as an alternative to the usual cooking method of deep-frying in an attempt to reduce the calorie or fat content of their snack products.

Baking Blind - Broiling - Flashbaking

Boiling

Boiling is the rapid vaporization of a liquid, which occurs when a liquid is heated to its boiling point, the temperature at which the vapor pressure of the liquid is equal to the pressure exerted on the liquid by the surrounding environmental pressure. While below the boiling point a liquid evaporates from its surface, at the boiling point vapor bubbles come from the bulk of the liquid. For this to be possible, the vapor pressure must be sufficiently high to win the atmospheric pressure, so that the bubbles can be "inflated". Thus, the difference between evaporation and boiling is "mechanical", rather than thermodynamical. The boiling point is lowered when the pressure of the surrounding atmosphere is reduced, for example by the use of a vacuum pump or at high altitudes. Boiling occurs in three characteristic stages, which are *nucleate*, *transition* and *film boiling*. These stages generally take place from low to high heating surface temperatures, respectively.

Boiling Stages

Nucleate Boiling

Nucleate boiling is characterized by the growth of bubbles on a heated surface, which rise from discrete points on a surface, whose temperature is only slightly above the liquid's. In general, the number of nucleation sites are increased by an increasing surface temperature.

An irregular surface of the boiling vessel (i.e. increased surface roughness) can create additional nucleation sites, while an exceptionally smooth surface, such as glass, lends itself to superheating. Under these conditions, a heated liquid may show boiling delay and the temperature may go somewhat above the boiling point without boiling.

Transition Boiling

Transition boiling may be defined as the unstable boiling, which occurs at surface temperatures between the maximum attainable in nucleate and the minimum attainable in film boiling.

The formation of bubbles in a heated liquid is a complex physical process which often involves cavitation and acoustic effects, such as the broad-spectrum hiss one hears in a kettle not yet heated to the point where bubbles boil to the surface.

Film Boiling

If a surface heating the liquid is significantly hotter than the liquid then film boiling will occur, where a thin layer of vapor, which has low thermal conductivity insulates the surface. This condition of a vapor film insulating the surface from the liquid characterizes *film boiling.*

Applications

Distillation

In distillation, boiling is used to separate mixtures. This is possible because the vapor rising from a boiling fluid generally has a ratio of components different from that in the liquid. It must be boiled for a full 10 minutes in order to kill all pathogens and be considered sterile.

Boiling for Water Sterilization

Boiling can be used as a method of water disinfection but is only advocated as an emergency water treatment method, or as a method of portable water purification in rural or wilderness settings without access to a potable water infrastructure. Although bringing water to the boil is effective in killing or inactivating most bacteria, viruses and pathogens, some are resistant and can survive several minutes boiling especially at high altitude where the temperature at which water boils is reduced.

Boiling in Cooking

In cooking, *boiling* is the method of cooking food in boiling water, or other water-based liquid such as stock or milk. Simmering is gentle boiling, while in poaching the cooking liquid moves but scarcely bubbles. Boiling is a very harsh technique of cooking. Delicate foods such as fish cannot be cooked in this fashion because the bubbles can damage the food. Foods such as red meat, chicken, and root vegetables can be cooked with this technique because of their tough texture.

The open-air boiling point of water is typically considered to be 100 °C or 212 °F. Pressure and a change in composition of the liquid may alter the boiling point of the liquid. For this reason, high elevation cooking generally takes longer since boiling point is a function of atmospheric pressure. In Denver, Colorado, which is at an elevation of about one mile, water boils at approximately 95 °C. Depending on the type of food and the elevation, the boiling water may not be hot enough to cook the food properly. Similarly, increasing the pressure as in a pressure cooker raises the temperature of the contents above the open air boiling point.

Adding a water soluble substance, such as salt or sugar also increases the boiling point. This is called boiling-point elevation. However, the effect is very small, and the boiling point will be increased by an insignificant amount. Due to variations in composition and pressure, the boiling point of water is almost never exactly 100 °C, but rather close enough for cooking.

Bringing water to a boil is generally done by applying maximal heat, then shutting off when the water has come to a boil, which is known as bang–bang control. Keeping water at or below a boil requires more careful control of temperature, particularly by using feedback.

In places where the available water supply is contaminated with disease-causing bacteria, boiling water and allowing it to cool before drinking it is practiced as a valuable health measure. Boiling water for a few minutes kills most bacteria, amoeba, and other microbial pathogens. It thus can help prevent cholera, dysentery, and other diseases caused by microorganisms. Foods suitable for boiling include vegetables, starchy foods such as rice, noodles and potatoes, eggs, meats, sauces, stocks and soups. Boiling has several advantages. It is safe and simple, and it is appropriate for large-scale cookery. Older, tougher, cheaper cuts of meat and poultry can be made digestible. Nutritious, well flavoured stock is produced. Also, maximum color and nutritive value is retained when cooking green vegetables, provided boiling time is kept to the minimum.

On the other hand, there are several disadvantages. There is a loss of soluble vitamins from foods to the water (if the water is discarded), and some boiled foods can look unattractive. Boiling can also be a slow method of cooking food.

Boiling can be done in several ways: The food can be placed into already rapidly boiling water and left to cook, the heat can be turned down and the food can be simmered; or the food can also be placed into the pot, and cold water may be added to the pot. This may then be boiled until the food is satisfactory. Water on the outside of a pot, i.e. a wet pot, increases the time it takes the pot of water to boil. The pot will heat at a normal rate once all excess water on the outside of the pot evaporates.

Levels of Boiling

In Chinese cuisine, particularly tea brewing, one distinguishes five stages of boiling: "*shrimp eyes,* the first tiny bubbles that start to appear on the surface of the kettle water, *crab eyes,* the secondary, larger bubbles, then *fish eyes,* followed by *rope of pearls,* and finally *raging torrent* [*rolling boil*]".

In detail:

shrimp eyes: about 70-80 °C (155–175 °F) – separate bubbles, rising to top.

crab eyes: about 80 °C (175 °F) – streams of bubbles.

fish eyes: about 80-90 °C (175–195 °F) – larger bubbles.

rope of pearls: about 90-95 °C (195–205 °F) – steady streams of large bubbles.

raging torrent: rolling boil, swirling and roiling.

Blanching - Braising - Coddling - Double steaming - Infusion - Poaching - Pressure cooking - Simmering - Steaming - Steeping - Stewing - Vacuum flask cooking.

Frying

Frying is the cooking of food in oil or fat, a technique that originated in ancient Egypt around 2500 BC. Chemically, oils and fats are the same, differing only in melting point, but the distinction is only made when needed. In commerce, many fats are called oils by custom, e.g. palm oil and coconut oil, which are solid at room temperature.

History

Frying is believed to have originated in ancient Egypt around 2500 BC.

Details

Fats can reach much higher temperatures than water at normal atmospheric pressure. Through frying, one can sear or even carbonize the surface of foods while caramelizing sugars. The food is cooked much more quickly and has a characteristic crispness and texture. Depending on the food, the fat will penetrate it to varying degrees, contributing richness, lubricity, and its own flavour.

Frying techniques vary in the amount of fat required, the cooking time, the type of cooking vessel required, and the manipulation of the food. Sautéing, stir frying, pan frying, shallow frying, and deep frying are all standard frying techniques.

Sautéing and stir-frying involve cooking foods in a thin layer of fat on a hot surface, such as a frying pan, griddle, wok, or sauteuse. Stir frying involves frying quickly at very high temperatures, requiring that the food be stirred continuously to prevent it from adhering to the cooking surface and burning.

Shallow frying is a type of pan frying using only enough fat to immerse approximately one-third to one-half of each piece of food; fat used in this technique is typically only used once.

Deep-frying, on the other hand, involves totally immersing the food in hot oil, which is normally topped up and used several times before being disposed. Deep-frying is typically a much more involved process, and may require specialized oils for optimal results.

Deep frying is now the basis of a very large and expanding worldwide industry. Fried products have consumer appeal in all age groups, and the process is quick, can easily be made continuous for mass production, and the food emerges sterile and dry, with a relatively long shelf life. The end products can then be easily packaged for storage and distribution. Examples are potato chips, french fries, nuts, doughnuts, instant noodles, etc.

Smoking

Smoking is the process of flavouring, cooking, or preserving food by exposing it to the smoke from burning or smoldering plant materials, most often wood. Meats and fish are the most common smoked foods, though cheeses, vegetables, and ingredients used to make beverages such as whisky, Rauchbier and *lapsang souchong* tea are also smoked.

In Europe, alder is the traditional smoking wood, but oak is more often used now, and beech to a lesser extent. In North America, hickory, mesquite, oak, pecan, alder, maple, and fruit-tree woods, such as apple, cherry and plum, are commonly used for smoking. Other fuels besides wood can also be employed, sometimes with the addition of flavouring ingredients. Chinese tea-smoking uses a mixture of uncooked rice, sugar, and tea, heated at the base of a wok. Some North American ham and bacon makers smoke their products over burning corncobs. Peat is burned to dry and smoke the barley malt used to make whisky and some beers. In New Zealand, sawdust from the native manuka (tea tree) is commonly used for hot smoking fish. In Iceland, dried sheep dung is used to cold smoke fish, lamb, mutton and whale, resulting in a unique and rather strongly smoked flavour.

Historically, farms in the western world included a small building termed the *smokehouse*, where meats could be smoked and stored. This was generally well-separated from other buildings both because of the fire danger and because of the smoke emanations.

Types

- Cold smoking can be used as a flavour enhancer for items such as chicken breasts, beef, pork chops, salmon, scallops, and steak. The item can be cold smoked for just long enough to give some flavour. Some cold smoked foods are baked, grilled, roasted, or sautéed before eating. Smokehouse temperatures for cold smoking are below 100 °F (38 °C). In this temperature range, foods take on a smoked flavour, but remain relatively moist. Cold smoking does not cook foods.
- Hot smoking exposes the foods to smoke and heat in a controlled environment. Although foods that have been hot smoked are often reheated or cooked, they are typically safe to eat without further cooking. Hams and ham hocks are fully cooked once they are properly smoked. Hot smoking occurs within the range of 165 °F (74 °C) to 185 °F (85 °C). Within this temperature range, foods are fully cooked, moist, and flavourful. If the smoker is allowed to get hotter than 185 °F (85 °C), the foods will shrink excessively, buckle, or even split. Smoking at high temperatures also reduces yield, as both moisture and fat are "cooked" away.
- Smoke roasting or smoke baking refers to any process that has the attributes of smoking combined with either roasting or baking. This smoking method is sometimes referred to as "barbecuing", "pit baking", or "pit roasting". It may be done in a smoke roaster, closed wood-fired masonry oven or barbecue pit, any smoker that can reach above 250 °F (121 °C), or in a conventional oven by placing a pan filled with hardwood chips on the floor of the oven so the chips smolder and produce a smokebath. However, this should only be done in a well-ventilated area to prevent carbon monoxide poisoning.

Wood Smoke

Hardwoods are made up mostly of three materials: cellulose, hemicellulose, and lignin. Cellulose and hemicellulose are the basic structural material of the wood cells; lignin acts as a kind of cell-bonding glue. Some softwoods, especially pines and firs, hold significant quantities of resin, which produces a harsh-tasting soot when burned; these woods are not often used for smoking.

Cellulose and hemicellulose are aggregate sugar molecules; when burnt, they effectively caramelize, producing carbonyls, which provide most of the color components and sweet, flowery, and fruity aromas.

Lignin, a highly complex arrangement of interlocked phenolic molecules, also produces a number of distinctive aromatic elements when burnt, including smoky, spicy, and pungent compounds such as guaiacol, phenol, and syringol, and sweeter scents such as the vanilla-scented vanillin and clove-like isoeugenol. Guaiacol is the phenolic compound most responsible for the "smokey" taste, while syringol is the primary contributor to smokey aroma. Wood also contains small quantities of proteins, which contribute roasted flavours. Many of the odor compounds in wood smoke, especially the phenolic compounds, are unstable, dissipating after a few weeks or months.

A number of wood smoke compounds act as preservatives. Phenol and other phenolic compounds in wood smoke are both antioxidants, which slow rancidification of animal fats, and antimicrobials, which slow bacterial growth. Other antimicrobials in wood smoke include formaldehyde, acetic acid, and other organic acids, which give wood smoke a low pH—about 2.5. Some of these compounds are toxic to people as well, and may have health effects in the quantities found in cooking applications.

Since different species of trees have different ratios of components, various types of wood do impart a different flavour to food. Another important factor is the temperature at which the wood burns. High-temperature fires see the flavour molecules broken down further into unpleasant or flavorless compounds.

The optimal conditions for smoke flavour are low, smoldering temperatures between 570 and 750 °F (299 and 399 °C). This is the temperature of the burning wood itself, not of the smoking environment, which uses much lower temperatures. Woods that are high in lignin content tend to burn hot; to keep them smoldering requires restricted oxygen supplies or a high moisture content. When smoking using wood chips or chunks, the combustion temperature is often raised by soaking the pieces in water before placing them on a fire.

Types of Smoker

There are a few basic types of smoker designs, each with their own advantages and disadvantages.

Charcoal Smokers

Offset Smokers

The main characteristics of the offset smoker are that the cooking chamber is usually cylindrical in shape, with a shorter, smaller diameter cylinder attached to the bottom of one end for a firebox. To cook the meat, a small fire is lit in the firebox, where airflow is tightly controlled.

The heat and smoke from the fire is drawn through a connecting pipe or opening into the cooking chamber. The heat and smoke cook and flavour the meat before escaping through an exhaust vent at the opposite end of the cooking chamber. Most manufacturers' models are based on this simple but effective design, and this is what most people picture when they think of a "BBQ smoker." Even large capacity commercial units use this same basic design of a separate, smaller fire box and a larger cooking chamber.

The UDS

The Upright Drum Smoker (also referred to as an Ugly Drum Smoker or UDS) is exactly what its name suggests; an upright steel drum that has been modified for the purpose of pseudo-indirect hot smoking. There are many ways to accomplish this, but the basics include the use of a complete steel drum, a basket to hold charcoal near the bottom, and cooking rack (or racks) near the top; all covered by a vented lid of some sort. They have been built using many different sizes of steel drums (30 gallon, 55 gallon, and 85 gallon for example), but the most popular size is the common 55 gallon drum.

This design is similar to smoking with indirect heat due to the distance from the coals and the racks (typically 24"). The temperatures used for smoking are controlled by limiting the amount of air intake at the bottom of the drum, and allowing a similar amount of exhaust out of vents in the lid. UDSs are very efficient with fuel consumption and flexible in their abilities to produce proper smoking conditions, with or without the use of a water pan or drip pan. Most UDS builders/users would say a water pan defeats the true pit BBQ nature of the UDS, as the drippings from the smoked meat should land on the coals, burning up, and imparting a unique flavour one cannot get with a water pan.

Vertical Water Smoker

A vertical water smoker (also referred to as a bullet smoker because of its shape) is a variation of the upright drum smoker. It uses charcoal or wood to generate smoke and heat, and contains a water bowl between the fire and the cooking grates. The water bowl serves to hold the temperature down and also to add humidity to the smoke chamber.

In addition, the bowl catches any drippings from the meat that may cause a flare up. Vertical water smokers are extremely temperature stable and require very little adjustment once the desired temperature has been reached. Because of their relatively low cost and stable temperature, they are sometimes used in barbecue competitions where propane and electric smokers are not allowed.

Propane Smoker

A propane smoker is designed to allow the smoking of meat in a somewhat more controlled environment. The primary differences are the sources of heat and of the smoke. In a propane smoker, the heat is generated by a gas burner directly under a steel or iron box containing the wood or charcoal that provides the smoke. The steel box has few vent holes, on the top of the box only. By starving the heated wood of oxygen, it smokes instead of burning. Any combination of woods and charcoal may used. This method uses less wood.

Smoke Box Method

This more traditional method uses a two box system: The fire box and the food box. The fire box is typically adjacent or under the cooking box, and can be controlled to a finer degree. The heat and smoke from the fire box exhausts into the food box, where it is used to cook and smoke the meat. These may be as simple as an electric heating element with a pan of wood chips placed on it, although more advanced models have finer temperature controls.

Commercial Smoke House

Commercial smokehouses, mostly made from stainless steel, have independent systems for smoke generation and cooking. Smoke generators use friction, an electric coil or a small flame to ignite sawdust on demand. Heat from steam coils or gas flames is balanced with live steam or water sprays to control the temperature and humidity. Elaborate air handling systems reduce hot or cold spots, to reduce variation in the finished product. Racks on wheels or rails are used to hold the product and facilitate movement.

Preservation

Smoke is an antimicrobial and antioxidant, but smoke alone is insufficient for preserving food in practice, unless combined with another preservation method. The main problem is the smoke compounds adhere only to the outer surfaces of the food; smoke does not actually penetrate far into meat or fish. In modern times, almost all smoking is carried out for its flavour. Artificial smoke flavouring can be purchased as a liquid to mimic the flavour of smoking, but not its preservative qualities.

In the past, smoking was a useful preservation tool, in combination with other techniques, most commonly salt-curing or drying. In some cases, particularly in climates without much hot sunshine, smoking was simply an unavoidable side effect of drying over a fire.

For some long-smoked foods, the smoking time also served to dry the food. Drying, curing, or other techniques can render the interior of foods inhospitable to bacterial life, while the smoking gives the vulnerable exterior surfaces an extra layer of protection.

For oily fish smoking is especially useful, as its antioxidant properties delay surface fat rancidification. (Interior fat is not as exposed to oxygen, which is what causes rancidity.) Some heavily-salted, long-smoked fish can keep without refrigeration for weeks or months. Such heavily-preserved foods usually require a treatment such as boiling in fresh water to make them palatable before eating.

Cancer Risk

"Of various sources of N-nitroso compounds, intake of smoked and salted fish was significantly (RR = 2.58, 95% CI 1.21 " 5.51) and intake of cured meat was non-significantly (RR = 1.84, 95% CI 0.98– 3.47) associated with risk of colorectal cancer."

Some Smoked Foods

Many foods can be smoked. Some of the more common are listed here.

Beverages

- *Lapsang souchong* tea leaves are smoked and dried over pine or cedar fires
- Malt beverages
  - o The malt used to make whisky
  - o Rauchbier (smoked beer).

Fruit

- Capsicums: chipotles (smoked, ripe jalapeño peppers), paprika
- Prunes (dried plums) can be smoked while drying
- *Wumei* are smoked ume fruit.

Protein

- Cheeses
- Fish :

  - o Eel popular in eastern/northern Europe and Japan, where it is called *unagi*
  - o Grimsby traditional smoked fish (cod and haddock)
  - o Haddock and Arbroath Smokies (haddock)
  - o Kippers and bloater (herring)
  - o Salmon.
- Beef :
  - o Traditionally-prepared jerky
  - o Pastrami (pickled, spiced and smoked beef brisket).
- Pork :
  - o Bacon
  - o Prosciutto
  - o Ham
  - o Various sausages.

Spices

- Paprika
- Salt.

Raw Food Preparation

Certain cultures highlight animal and vegetable foods in their raw state. Salads consisting of raw vegetables or fruits are common in many cuisines. Sashimi in Japanese cuisine consists of raw sliced fish or other meat, and sushi often incorporates raw fish or seafood. Steak tartare and salmon tartare are dishes made from diced or ground raw beef or salmon, mixed with various ingredients and served with baguettes, brioche, or frites. In Italy, carpaccio is a dish of very thinly sliced raw beef, drizzled with a vinaigrette made with olive oil. The health food movement known as raw foodism promotes a mostly vegan diet of raw fruits, vegetables, and grains prepared in various ways, including juicing, food dehydration, sprouting, and other methods of preparation that do not heat the food above 118 °F (47.8 °C).

A ceviche is a Latin American dish made with raw meat that is "cooked" from the highly acidic citric juice from lemons and limes along with other aromatics such as garlic.

Restaurants

Restaurants employ trained chefs who prepare food, and trained waitstaff to serve the customers. The term restaurant is credited to the French from the 19th century, as it relates to the restorative nature of the bouillons that were once served in them. However, the concept pre-dates the naming of these establishments, as evidence suggests commercial food preparation may have existed during the age of the city of Pompeii, and urban sales of prepared foods may have existed in China during the Song Dynasty. The coffee shops or cafés of 17th century Europe may also be considered an early version of the restaurant. In 2005, the population of the United States spent $496 billion for out-of-home dining. Expenditures by type of out-of-home dining were as follows: 40% in full-service restaurants, 37.2% in limited service restaurants (fast food), 6.6% in schools or colleges, 5.4% in bars and vending machines, 4.7% in hotels and motels, 4.0% in recreational places, and 2.2% in others, which includes military bases.

Food Manufacture

Food manufacture is the process by which food is manufactured.

History

Early food processing techniques were limited by the available food preservation, packaging and transportation. Early food processing mainly involved salting, curing, curdling, drying, pickling and smoking. An example of an early processed food product is cheese.

During the industrialisation era in the 19th century, food manufacturing arose. This development took advantage of new mass markets and emerging new technology, such as milling, preservation, packaging and labelling and transportation. It brought the advantages of prepared time-saving food to the bulk of ordinary people who did not employ domestic servants.

Industry Organization

At the start of the 21st century, a two-tier structure has arisen, with a few international food processing giants controlling a wide range of well-known food brands, and a populous number of small local or national food processing companies. Packaged foods are manufactured outside the home for purchase. This can be as simple as a butcher preparing meat, or as complex as a modern international food industry. Early food processing techniques were limited by available food preservation, packaging, and transportation. This mainly involved salting, curing, curdling, drying, pickling, fermenting, and smoking. Food manufacturing arose during the industrial revolution in the 19th century. This development took advantage of new mass markets and emerging new technology, such as milling, preservation, packaging and labelling, and transportation. It brought the advantages of pre-prepared time-saving food to the bulk of ordinary people who did not employ domestic servants.

At the start of the 21st century, a two-tier structure has arisen, with a few international food processing giants controlling a wide range of well-known food brands. There also exists a wide array of small local or national food processing companies. Advanced technologies have also come to change food manufacture. Computer-based control systems, sophisticated processing and packaging methods, and logistics and distribution advances can enhance product quality, improve food safety, and reduce costs.

Production Processes

Extrusion is the forcing or thrusting out of an object. Within most industries, extrusion is used in tandem with molds and cutting mechanisms to create uniform, identically-shaped and formed objects.

Within the food industry, contract extrusion can have many applications. It can be used in the production and packaging of cereals, noodles, cookie dough, and many other items commonly found in grocery stores. Twin-screw extruders are used to grind and mix ingredients before extruding them through a mold to the desired shape.

On a twin-screw extruder, an engine powers two large, screw-shaped devices that rotate opposite one another. The distance between the two screws determines the size to which the ingredients are ground. The ingredients can also be heated or cooled throughout the food extrusion process.

At the start of the extrusion process, raw ingredients are fed into the primary feed port. In nearly all cases, these are dry ingredients. Immediately after being fed into the machine, the ingredients are crushed and ground into ideally sized particles. In complex mixtures, an extruder may have multiple ports along its sides. These ports allow for mid-process additions of other ingredients. The more ports, the more ingredients can be added into the mixture. This allows for the gradual combination of ingredients, to ensure ideal composition and integrity of the product. Twin-screw extruders are used primarily for mixtures, but single-components can be run through the extruder to achieve the ideal consistency and final product.

Now that the ingredients have been mixed, they can be heated, cooled, dried, or added into another mixture. The ingredients or products, having been sufficiently ground and combined, move into the actual extrusion process. In this step of the process, they are forced through a die to give the product the desired shape. Think of it like pushing play dough through a mold. These die can be shaped in an incredible array of options, and can create hollow spaces as well as solid products. Once pushed through the die, the shaped product is shorn off, and the extrusion process is complete. Depending on the product, the sheering can be rapid, for when the product is small, or slow, for when more of the product must be extruded before cutting it off.

One benefit of the extrusion process is in the quality and consistency of the finished product. It also allows you to replicate the exact product endlessly, so long as you continue to feed the ingredients into the extruder.

As you may suspect, few manufacturing and packaging facilities have this capability in-house. Contract extrusion, then, is the process by which one company or purchaser utilizes an outside company to complete their extrusion for them. Depending on many elements of the business, industry and specific technical requirements, contract extrusion and outsourcing the extrusion process can save a company money, time, and manpower.

Commercial Trade

International Exports and Imports

The World Bank reported that the European Union was the top food importer in 2005, followed at a distance by the USA and Japan. Food is now traded and marketed on a global basis. The variety and availability of food is no longer restricted by the diversity of locally grown food or the limitations of the local growing season. Between 1961 and 1999, there was a 400% increase in worldwide food exports. Some countries are now economically dependent on food exports, which in some cases account for over 80% of all exports.

In 1994, over 100 countries became signatories to the Uruguay Round of the General Agreement on Tariffs and Trade in a dramatic increase in trade liberalization. This included an agreement to reduce subsidies paid to farmers, underpinned by the WTO enforcement of agricultural subsidy, tariffs, import quotas, and settlement of trade disputes that cannot be bilaterally resolved. Where trade barriers are raised on the disputed grounds of public health and safety, the WTO refer the dispute to the Codex Alimentarius Commission, which was founded in 1962 by the United Nations Food and Agriculture Organization and the World Health Organization. Trade liberalization has greatly affected world food trade.

Marketing and Retailing

Food marketing brings together the producer and the consumer. It is the chain of activities that brings food from "farm gate to plate". The marketing of even a single food product can be a complicated process involving many producers and companies. For example, fifty-six companies are involved in making one can of chicken noodle soup. These businesses include not only chicken and vegetable processors but also the companies that transport the ingredients and those who print labels and manufacture cans. The food marketing system is the largest direct and indirect non-government employer in the United States.

In the pre-modern era, the sale of surplus food took place once a week when farmers took their wares on market day into the local village marketplace. Here food was sold to grocers for sale in their local shops for purchase by local consumers. With the onset of industrialization and the development of the food processing industry, a wider range of food could be sold and distributed in distant locations. Typically early grocery shops would be counter-based shops, in which purchasers told the shop-keeper what they wanted, so that the shop-keeper could get it for them.

In the 20th century, supermarkets were born. Supermarkets brought with them a self service approach to shopping using shopping carts, and were able to offer quality food at lower cost through economies of scale and reduced staffing costs. In the latter part of the 20th century, this has been further revolutionized by the development of vast warehouse-sized, out-of-town supermarkets, selling a wide range of food from around the world. Unlike food processors, food retailing is a two-tier market in which a small number of very large companies control a large proportion of supermarkets. The supermarket giants wield great purchasing power over farmers and processors, and strong influence over consumers. Nevertheless, less than 10% of consumer spending on food goes to farmers, with larger percentages going to advertising, transportation, and intermediate corporations.

Prices

It was reported on March 24, 2008, that consumers worldwide faced rising food prices. Reasons for this development include changes in the weather and dramatic changes in the global economy, including higher oil prices, lower food reserves, and growing consumer demand in China and India. In the long term, prices are expected to stabilize. Farmers will grow more grain for both fuel and food and eventually bring prices down. Already this is happening with wheat, with more crops to be planted in the United States, Canada, and Europe in 2009. However, the Food and Agriculture Organization projects that consumers still have to deal with more expensive food until at least 2018.

It is rare for the spikes to hit all major foods in most countries at once. Food prices rose 4% in the United States in 2007, the highest rise since 1990, and are expected to climb as much again in 2008. As of December 2007, 37 countries faced food crises, and 20 had imposed some sort of food-price controls. In China, the price of pork jumped 58% in 2007. In the 1980s and 1990s, farm subsidies and support programs allowed major grain exporting countries to hold large surpluses, which could be tapped during food shortages to keep prices down. However, new trade policies have made agricultural production much more responsive to market demands, putting global food reserves at their lowest since 1983.

Food prices are rising, wealthier Asian consumers are westernizing their diets, and farmers and nations of the third world are struggling to keep up the pace. The past five years have seen rapid growth in the contribution of Asian nations to the global fluid and powdered milk manufacturing industry, which in 2008 accounted for more than 30% of production, while China alone accounts for more than 10% of both production and consumption in the global fruit and vegetable processing and preserving industry.

The trend is similarly evident in industries such as soft drink and bottled water manufacturing, as well as global cocoa, chocolate, and sugar confectionery manufacturing, forecast to grow by 5.7% and 10.0% respectively during 2008 in response to soaring demand in Chinese and Southeast Asian markets.

Famine and Hunger

Food deprivation leads to malnutrition and ultimately starvation. This is often connected with famine, which involves the absence of food in entire communities. This can have a devastating and widespread effect on human health and mortality. Rationing is sometimes used to distribute food in times of shortage, most notably during times of war.

Starvation is a significant international problem. Approximately 815 million people are undernourished, and over 16,000 children die per day from hunger-related causes. Food deprivation is regarded as a deficit need in Maslow's hierarchy of needs and is measured using famine scales.

Food Aid

Food aid can benefit people suffering from a shortage of food. It can be used to improve peoples' lives in the short term, so that a society can increase its standard of living to the point that food aid is no longer required. Conversely, badly managed food aid can create problems by disrupting local markets, depressing crop prices, and discouraging food production. Sometimes a cycle of food aid dependence can develop. Its provision, or threatened withdrawal, is sometimes used as a political tool to influence the policies of the destination country, a strategy known as food politics. Sometimes, food aid provisions will require certain types of food be purchased from certain sellers, and food aid can be misused to enhance the markets of donor countries. International efforts to distribute food to the neediest countries are often coordinated by the World Food Programme.

Safety

Foodborne illness, commonly called "food poisoning", is caused by bacteria, toxins, viruses, parasites, and prions. Roughly 7 million people die of food poisoning each year, with about 10 times as many suffering from a non-fatal version. The two most common factors leading to cases of bacterial foodborne illness are cross-contamination of ready-to-eat food from other uncooked foods and improper temperature control. Less commonly, acute adverse reactions can also occur if chemical contamination of food occurs, for example from improper storage, or use of non-food grade soaps and disinfectants. Food can also be adulterated by a very wide range of articles (known as "foreign bodies") during farming, manufacture, cooking, packaging, distribution, or sale. These foreign bodies can include pests or their droppings, hairs, cigarette butts, wood chips, and all manner of other contaminants. It is possible for certain types of food to become contaminated if stored or presented in an unsafe container, such as a ceramic pot with lead-based glaze.

Food poisoning has been recognized as a disease since as early as Hippocrates. The sale of rancid, contaminated, or adulterated food was commonplace until the introduction of hygiene, refrigeration, and vermin controls in the 19th century. Discovery of techniques for killing bacteria using heat, and other microbiological studies by scientists such as Louis Pasteur, contributed to the modern sanitation standards that are ubiquitous in developed nations today. This was further underpinned by the work of Justus von Liebig, which led to the development of modern food storage and food preservation methods.

In more recent years, a greater understanding of the causes of food-borne illnesses has led to the development of more systematic approaches such as the Hazard Analysis and Critical Control Points (HACCP), which can identify and eliminate many risks.

Recommended measures for ensuring food safety include maintaining a clean preparation area with foods of different types kept separate, ensuring an adequate cooking temperature, and refrigerating foods promptly after cooking.

Foods that spoil easily, such as meats, dairy, and seafood, must be prepared a certain way to avoid contaminating the people for whom they are prepared. As such, the general rule of thumb is that cold foods (such as dairy products) should be kept cold and hot foods (such as soup) should be kept hot until storage. Cold meats, such as chicken, that are to be cooked should not be placed at room temperature for thawing, at the risk of dangerous bacterial growth, such as *Salmonella* or *E. coli*.

Allergies

Some people have allergies or sensitivities to foods which are not problematic to most people. This occurs when a person's immune system mistakes a certain food protein for a harmful foreign agent and attacks it. About 2% of adults and 8% of children have a food allergy.

The amount of the food substance required to provoke a reaction in a particularly susceptible individual can be quite small. In some instances, traces of food in the air, too minute to be perceived through smell, have been known to provoke lethal reactions in extremely sensitive individuals.

Common food allergens are gluten, corn, shellfish (mollusks), peanuts, and soy. Allergens frequently produce symptoms such as diarrhea, rashes, bloating, vomiting, and regurgitation. The digestive complaints usually develop within half an hour of ingesting the allergen. Rarely, food allergies can lead to a medical emergency, such as anaphylactic shock, hypotension (low blood pressure), and loss of consciousness. An allergen associated with this type of reaction is peanut, although latex products can induce similar reactions. Initial treatment is with epinephrine (adrenaline), often carried by known patients in the form of an Epi-pen or Twinject.

Diet

In nutrition, diet is the sum of food consumed by a person or other organism. Dietary habits are the habitual decisions an individual or culture makes when choosing what foods to eat. With the word diet, it is often implied the use of specific intake of nutrition for health or weight-management reasons (with the two often being related). Although humans are omnivores, each culture and each person holds some food preferences or some food taboos, due to personal tastes or due to ethical reasons. Individual dietary choices may be more or less healthful. Proper nutrition requires the proper ingestion and, equally important, the absorption of vitamins, minerals, and food energy in the form of carbohydrates, proteins, and fats. Dietary habits and choices play a significant role in health and mortality, and can also define cultures and play a role in religion.

Traditional Diets

Traditional diets are those of native populations such as the Native Americans, Khoisan or Australian Aborigines. Often, to qualify for cultural cuisine, traditional diets include more organic farming and seasonal food according to food origins.

Traditional diets vary with availability of local resources, such as fish in coastal towns, eels and eggs in estuary settlements, or squash, corn and beans in farming towns, as well as with cultural and religious customs and taboos. In some cases, the crops and domestic animals that characterize a traditional diet have been replaced by modern high-yield crops, and are no longer available. The slow food movement attempts to counter this trend and to preserve traditional diets.

A recent study has suggested that traditional diets may have been more balanced than first thought. New research indicates grains were part of the diet of ancient people in Italy, Russia and the Czech Republic.

Religious and Cultural Dietary Choices

Some cultures and religions have restrictions concerning what foods are acceptable in their diet. For example, only Kosher foods are permitted by Judaism, and Halal foods by Islam.

Diet and Life Outcomes

A three-decade long study published in the British medical journal, The Lancet, found that Guatemalan men who had been well-fed soon after they were born earned almost 50% more in average salary than those who had not. The blind trial was performed by giving a high-nutrition supplement to some infants and a lower-nutrition supplement to others, with only the researchers knowing which infants received which supplements. The infants that received the high-nutrition supplement had higher average salaries as adults.

Individual Dietary Choices

Writers such as Michael Pollan and Mark Bittman urge reduced animal consumption in the developed world for improved health and reduced impact on the environment. Many people choose to forgo food from animal sources to varying degrees (flexitarianism, vegetarianism, veganism, fruitarianism) for health reasons, or issues surrounding morality, or to reduce their personal impact on the environment. Raw foodism is another contemporary trend. These diets may require tuning or supplementation to meet ordinary nutritional needs.

Diets for Weight Management

A particular diet may be chosen to seek weight loss or weight gain. Changing a subject's dietary intake, or "going on a diet", can change the energy balance and increase or decrease the amount of fat stored by the body. Some foods are specifically recommended, or even altered, for conformity to the requirements of a particular diet. These diets are often recommended in conjunction with exercise. Specific weight loss programs can be harmful to health, while others may be beneficial (and can thus be coined as healthy diets). The terms healthy diet and diet for weight management are often related, as the two promote healthy weight management.

Health

A healthy diet is one that is arrived at with the intent of improving or maintaining optimal health. This usually involves consuming nutrients by eating the appropriate amounts from all of the food groups, including an adequate amount of water. Since human nutrition is complex, a healthy diet may vary widely, and is subject to an individual's genetic make-up, environment, and health. For around 20% of the human population, lack of food and malnutrition are the main impediments to healthy eating. Conversely, people in developed countries have the opposite problem; concern is not about volume of food but appropriate choices.

Chapter 4

Enzymes in Food Processing

Food processing is a procedure in which food is prepared for consumption. People often use this term to refer specifically to making packaged foods, but technically anything which transforms raw ingredients into something else is a form of food processing, ranging from grilling vegetables in the back yard to making television dinners in a food manufacturing facility. The food processing sector employs large numbers of people, many of whom are unskilled labourers.

There are several purposes to food processing. The most basic goal is to prepare food which is palatable. This can include processing ingredients which are not safe to eat raw, flavouring foods to make them more interesting, and making dishes which comply with cultural and religions norms surrounding food, in addition to addressing issues such as allergies. Food processing is also usually intended to make food which is nutritious, and can include activities such as food fortification, in which vitamins and minerals are added to food during processing to increase the nutritional value.

Safety is also a major concern in food processing, especially industrial food processing to create packaged foods which are sold commercially. These facilities can be easily contaminated and the contamination can quickly spread, causing widespread illness. Part of making food safe includes processing it to remove any potential risks, such as bacteria in milk, in addition to maintaining strict safety procedures to reduce the risk of introducing harmful organisms during food processing.

In industrial food processing, these needs must be balanced with the need for preservation. Food which will be eaten in a few days needs to be stabilized so that it will retain texture and flavour in addition to staying safe. Other foods intended to be kept in dry storage for months or years must be specially processed so that they will not go bad. Some of the earliest forms of processing such as pickling and preserving foods echo the need to preserve foods effectively and safely which has challenged human cultures for thousands of years.

Many innovative techniques have been developed for food processing around the world to bring food to market while keeping it safe and flavourful. Certain processes are protected by patent and used only by people licensed to use the patent, such as the process behind Pringles® potato products, while others are widely used by everyone from home bakers to companies which produce diet meals on order for hospitals and institutions.

Food and feed is possibly the area where processing anchored in biological agents has the deepest roots. Despite this, process improvement or design and implementation of novel approaches has been consistently performed, and more so in recent years, where significant advances in enzyme engineering and biocatalyst design have fastened the pace of such developments. This chapter aims to provide an updated and succinct overview on the applications of enzymes in the food sector, and of progresses made, namely, within the scope of tapping for more efficient biocatalysts, through screening, structural modification, and immobilization of enzymes.

Targeted improvements aim at enzymes with enhanced thermal and operational stability, improved specific activity, modification of pH-activity profiles, and increased product specificity, among others. This has been mostly achieved through protein engineering and enzyme immobilization, along with improvements in screening. The latter has been considerably improved due to the implementation of high-throughput techniques, and due to developments in protein expression and microbial cell culture. Expanding screening to relatively unexplored environments (marine, temperature extreme environments) has also contributed to the identification and development of more efficient biocatalysts. Technological aspects are considered, but economic aspects are also briefly addressed.

Enzymes in Food Processing: A Condensed Overview on Strategies for Better Biocatalysts

Food processing through the use of biological agents is historically a well-established approach. The earliest applications go back to 6,000 BC or earlier, with the brewing of beer, bread baking, and cheese and wine making, whereas the first purposeful microbial oxidation dates from 2,000 BC, with vinegar production. Coming to modern days, in the late XIX, century Christian Hansen reported the use of rennet (a mixture of chymosin and pepsin) for cheese making, and production of bacterial amylases was started at Takamine (latter to become part of Genencor).

Pectinases were used for juice clarification in the 1930s, and for a short period during World War II, invertase was also used for the production of invert sugar syrup in a process that pioneered the use of immobilized enzymes in the sugar industry. Still, the large-scale application of enzymes only became really established in the 1960s, when the traditional acid hydrolysis of starch was replaced by an approach based in the use of amylases and amyloglucosidases (glucoamylases), a cocktail that some years latter would include glucose (xylose) isomerase.

From then on, the trend for the design and implementation of processes and production of goods anchored in the use of enzymes has steadily increased. Enzymes are currently among the well established products in biotechnology, from US $1.3 billion in 2002 to US $4 billion in 2007; it is expected to have reached US $5.1 billion in a rough 2009 year, and is anticipated to reach $7 billion by 2013. In the overall, this pattern corresponds to a rise in global demand slightly exceeding 6% yearly. Part of this market is ascribed to enzymes used in large-scale applications, among them are those used in food and feed applications.

These include enzymes used in baking, beverages and brewing, dairy, dietary supplements, as well as fats and oils, and they have typically been dominating one, only bested by the segment assigned to technical enzymes.

The latter includes enzymes in the detergent, personal care, leather, textile and pulp, and paper industries. A recent survey on world sales of enzymes ascribes 31% for food enzymes, 6% for feed enzymes and the remaining for technical enzymes. A relatively large number of companies are involved in enzyme manufacture, but major players are located in Europe, USA and Japan. Denmark is dominating, with Novozymes (45%) and Danisco (17%), moreover after the latter taking over Genencor (USA), with DSM (The Netherlands) and BASF (Germany) lagging behind, with 5% and 4%. The pace of development in emerging markets is suggestive that companies from India and China can join this restricted party in a very near future.

Relevant Enzymes: Tapping for Improved Biocatalysts

General Aspects and the Screening Approach

Roughly all classes of enzymes have an application within the food and feed area, but hydrolases are possibly the prevalent one. The widespread use of enzymes for food and feed processing is easily understandable, given their unsurpassed specificity, ability to operate under mild conditions of pH, temperature and pressure while displaying high activity and turnover numbers, and high biodegradability. Enzymes are furthermore generally considered a natural product. The whole contributes for developing sustainable and environmentally friendly processes, since there is a low amount of by-products, hence reducing the need for complex downstream process operations, and the energy requirements are relatively low. Life-cycle assessment (LCA) has confirmed, that within the range of given practical case studies, including food and feed processing, the implementation of enzyme-based technology has a positive impact on the environment. LCA is a methodology used to compare the environmental impact of alternative production technologies while providing the same user benefits.

Some of the broad generalizations on the limitations of enzymes for application as biocatalysts in commercial scale, namely, their high cost, low productivity and stability, and narrow range of substrates, have been rebutted. Aiming at improving the performance of biocatalysts for food and feed applications, particular care has been given to increasing thermal stability, enhancing the range of pH with catalytic activity and decreasing metal ions requirements, as well as to overcoming the susceptibility to typical inhibitory molecules.

Some examples of strategies taken to improve the performance of relevant enzymes for food and feed. Along with these different strategies focused on the enzyme molecule (namely, protein engineering, enzyme immobilization), the developments in recombinant DNA technology that occurred in the 1980s also had a huge impact on the application of enzymes in food and feed.

By allowing gene cloning in microorganisms compatible with industrial requirements, this methodology enabled cost-feasible production of enzymes that were naturally produced in conditions that prevented large-scale application (namely, enzymes from plant or animal cells, such as transglutaminase or even slow-growing microorganisms). When successfully implemented, the undertaken approaches allow: (a) continuous operations at relatively high temperatures; (b) eased implementation of enzyme cascade, given the reduced need for processing the reaction media (pH adjustments; metal ion removal/addition) throughout the intermediate steps of a multistep biotransformation (namely, starch to high fructose syrup); and (c) the use of raw substrates, preferably as high-concentrated solutions, hence cutting back in costs related to upstream processing and increasing productivity. Methodologies with a high level of parallelization, anchored in computer-monitored microtiter plates equipped with optic fibers and temperature control have also been developed. These provide high-throughput capability for a speedy and detailed characterization of the performance of enzymes. Particular focus was given to the prediction of the long-term stability of enzymes under moderate conditions using short-term runs (up to 3 hours).

One of the methodologies to obtain improved biocatalyst relies on in-vitro modifications, which will be addressed latter in this paper; another approach relies on screening efforts, which has been consistently undertaken, as summarized recently. Some focus is given to extremophiles, particularly thermophiles, since operation at high temperatures (roughly above 45–50°C) minimizes the risk of microbial contamination, a particularly delicate matter under continuous operation.

Furthermore, the extension of some reactions in relevant food applications is favored at relatively high temperatures (namely, isomerization of glucose to fructose), although care should be taken to avoid an operational environment that may lead by-product formation (namely, Maillard reactions). Examples of screened enzymes include the isolation of amylases, with some of them being calcium independent; amylopullulanases; fructosyltransferases; glucoamylases; glucose (xylose) isomerases; glucosidases; inulinases; levansucrases; pullulanases; and xylanases. Other examples of these enzymes, with some of which able to retain stability under temperatures of 90°C or higher, were reviewed by Gomes and Steiner. The majority of enzymes used in food and feed processing is of terrestrial microbial origin, and screening-efforts for isolation of promising enzyme-producing strains have accordingly been performed in such background. From some years now, marine environment has also been tapped as a source for useful enzymes from either microbial or higher organisms origin.

This latter environment has allowed the isolation of some promising biocatalysts, such as the heat-stable invertase/inulinase from Thermotoga neapolitana DSM 4359 or inulinase from Cryptococcus aureus, amylolytic enzymes, glucosidases and proteases from severalgenera, esterase from Vibrio fischeri, and glycosyl hydrolases.

Other examples of useful enzymes for food and feed, but isolated from higher organisms, are given in Operation at low temperatures is also welcome since it also reduces the risk of microbial contamination, enables some processes to be carried out with minimal deterioration of the raw material. These include protein processing, such as cheese maturing and milk coagulation with proteases; milk processing with lactase for lactose-free milk; clarification of fruit juices with pectinases to produce clear juice; or production of oligosaccharides.

Since extremophiles are often difficult to grow under typical laboratory conditions if not nonculturable at all, different approaches have been developed in order to assess the potential of enzymes from such microorganisms. One approach relies on the generation and screening of target genes from DNA libraries, which can be obtained from mixed microbial population from environmental samples.

Recombinant microorganisms can then be obtained using mesophiles as hosts where the genes of interest from extremophiles have been expressed. In order to screen the huge number of DNA-libraries typically generated for the intended property, high-throughput methods have been implemented. These methods are also widely used when protein engineering is carried out. This will be addressed in the following section.

Several enzymes (namely, α-amylases; pullulanases) currently used in food processing, namely, in starch hydrolysis, are actually produced by recombinant microorganisms. Despite some complexity in the implementation of their use in large-scale applications, partly resulting from lack of uniformity in the US and EU legislation, quite a few enzyme preparations have been accepted for industrial use.

Improving Biocatalysts: Beyond Screening

Taking advantage of the knowledge gathered on molecular biology, high-throughput processing, and computer-assisted design of proteins, in-vitro improvement of biocatalysts have been consistently implemented. Some of the research efforts in this area has focused on the biochemical and molecular mechanisms underlying the stability of enzymes from extremophiles. Such knowledge is also particularly useful for protein engineering of known enzymes, aiming at enhancing stability without compromising catalytic activity. Enhancing the stability of enzymes is of paramount importance when implementation of industrial processes is foreseen, since it allows for reducing the amount of enzyme used in the process.

Given that thermostability is determined by a series of short- and long-range interactions, it can be improved by several substitutions of amino acids in a single mutant, where the combination of each individual effect is usually roughly additive. The targeted improvements have not been restricted to thermostability, but they have also addressed other features, such as broadening the range of pH where the enzyme is active, or lessening the temperature of operation while retaining high activity.

Two methodologies can be used for protein engineering. The first is directed evolution of enzymes, through random mutagenesis and recombination, where the environmental adaptation is reproduced in-vitro in a much hastened timescale, towards the optimization of the intended property. In order to control the pathway of the process, either a screening test for the assessed feature is performed after each round of modification, or selective pressure is applied.

This methodology, which allows for a high throughput, has been extensively applied, aiming for more efficient biocatalysts. Some relevant examples in the area of food and feed processing include the following.

The first is the enhancement of the activity of the hyperthermostable glucose (xylose) isomerase from Thermotoga neapolitana at relatively low temperature and pH, without decay in thermostability. The enzyme from the parent strain is highly active at 97°C, but it retains only 10% of its activity at 60°C, and requires neutral pH for optimal activity. This pattern is often reported when glucose isomerases from hyperthermophilic strains operate in mesophilic environments.

Large-scale glucose isomerization is carried out at 55–60°C and slightly alkaline pH. This set of conditions results from the optimal range of pH (typically 7.0 to 9.0) and temperature (60 to 80°C) for glucose isomerization displayed by most of the glucose isomerases used, combined with process boundary conditions. The latter result from by-product and color formation occurring when the reaction is carried out at alkaline pH and high temperatures. There is therefore interest in selecting an enzyme able to operate efficiently at temperatures close to those currently used but at a lower pH.

The mutant glucose isomerase 1F1 obtained by Sriprapundh and coworkers displayed a roughly 5-fold higher activity at 60°C and pH 5.5, when compared with the parent T. neapolitana isomerase, and was more thermostable than the wild type isomerase. The activation energy required by the triple 1F1 mutant (V185T/L282P/F186S) was roughly half of the wild-type, hence allowing for high activity at relatively low temperatures. The encouraging results obtained suggest the soundness of the approach to obtain a mutant glucose isomerase competitive with those currently used, while being able to operate in a slightly acidic environment and 60°C. The second is the enhancement of the thermostability of the maltogenic amylase from Thermus sp. IM6501, of the amylosucrase from Neisseria polysaccharea, of the glucoamylase from Aspergillus niger, of a phytase from Escherichia coli, and of a xylanase from Bacillus subtilis.

Amylases and glucoamylases are enzymes used in starch processing, which involves temperatures typically in excess of 60°C; hence, improving thermal stability without decreasing enzyme activity is of relevance. Starch liquefaction is performed at 105°C in the presence of α-amylase, upon which the effluent reaction stream has to be cooled to 60°C, so that glucoamylases can be used.

In order to avoid, or at least minimize, the cooling step, thermostable glucoamylases are aimed at. Wang and coworkers obtained a multiply-mutated enzyme (N20C, A27C, S30P, T62A, S119P, G137A, T290A, H391Y), which displayed a 5.12 kJmol-1 increase in the free energy of thermal inactivation, when compared to the wild type, thus resulting in the enhanced thermal stability of the mutant.

Furthermore specific activities and catalytic efficiencies remained unaltered, when mutant and wild type were compared. Kim and coworkers obtained also a multiply-mutated amylase (R26Q, S169N, I333V, M375T, A398V, Q411L, P453L) which displayed an optimal reaction temperature 15°C higher than that of the wild-type and a half-life of roughly 170 min at 80°C, a temperature at which the wild-type ThMA was fully inactivated in less than 1 minute.

However, one of the mutations most accountable for enhanced thermal stability, M375T, close to the active site, also led to a 23% decrease in specific activity, as compared to the wild type. The amylosucrase engineered by Emond and coworkers was a double mutant (R20C/A451T), displaying a 10-fold increase in the half-life at 50°C compared to the wild-type enzyme. Actually, the mutant was claimed to be the only amylosucrase usable at 50°C. At the latter temperature, the mutant enabled the synthesis of amylose chains twice as long as those obtained by the wild-type enzyme at 30°C, for sucrose concentrations of 600 mM.

The mutant thus allowed for a process with increased yield in amylose chains, lower risk of contamination, enhanced substrate and product solubility and overall productivity. Phytases are added to animal feeds to improve phosphorus nutrition and to reduce phosphorus excretion, by promoting the hydrolysis of phytate into myoinositol and inorganic phosphate. Thermal stable enzymes are needed, since feed pelleting is carried out at high temperature (60 to 80°C).

Phytases produced by thermophiles do not provide a suitable approach, since they have low activity at the physiological temperature of animals. E. coli phytases, which are appealing to industrial application, due to the acidic pH optimum, specificity phytate, and resistance to pepsin digestion, were thus engineered in order to improve their thermal stability, without compromising the kinetic parameters. As a result, mutants were obtained, with roughly 20% increased thermostability at 80°C improved overall catalytic efficiency (kcat, turnover number/KM, Michaelis constant) within 50 to 150%, as compared to the wild type. No significant changes in the pH activity profile were observed, but for some mutants, containing a K46E substitution, that displayed a decrease in activity at pH 5.0.

Xylanases catalyze the cleavage of β1,4 bonds in xylan polymers. Accordingly, these enzymes can be used in dough making, in baking, in brewing and in animal feed compositions. When the latter contain cereals (namely, barley, maize, rye or wheat), or cereal by-products, xylanases improve the break-down of plant cell walls, which favours the ingestion of plant nutrients by the animals and consequently enhances feed consumption and growth rate.

Furthermore, the use of xylanases decreases the viscosity of xylan-containing feeds. As referred for phytases, the formulation of commercial feed often involves steps at high temperatures. Xylanases added to the formulations hence have to withstand these conditions, while they are to display high activity at about 40°C, which is the temperature in the intestine of animals.

However, most xylanases are inactive at temperatures exceeding 60°C, hence the need for enhancing thermal stability. Miyazaki and coworkers obtained a triple-mutant xylanase (Q7H, N8F, and S179C) which retained full activity for 2 hours at 60°C, whereas the wild-type enzyme was inactivated within 5 minutes under the same conditions.

The mutation also led to a 10°C increase in the optimal temperature for reaction and enhanced activity at higher temperatures, albeit at the cost of decreased activity at lower temperatures, as compared to the wild-type enzyme. Third is the enhancement of the activity of the amylosucrase from Neisseria polysaccharea.

Amylosucrases can be used for the modification or synthesis of amylose-type polymers from sucrose, but their industrial application is somehow thwarted by the low catalytic efficiency on sucrose and by side reactions leading to the formation of sucrose isomers. Van der Veen and co-works engineered mutant enzymes through error-prone PCR that displayed increases in activity up to 5-fold and in overall catalytic efficiency up to 2-fold, when compared to the wild-type enzyme.

Furthermore, the mutants were able to produce amylose polymers from 10 mM sucrose on, unlike the wild-type enzyme. Their work provides an illustrative example on the use of random mutagenesis and recombination for the enhancement of the catalytic properties of enzymes with application on food and feed. Another example was provided by Tian and coworkers who engineered a phytase from Aspergillus niger 113 through gene shuffling, to obtain mutants with enhanced catalytic properties. Hence, K41E and E121F substitutions allowed for increases in the specific activity of 2.5- and 3.1-fold, and of affinity for sodium phytate, as expressed by decreases in KM of roughly 35% and 25%, as compared to the wild-type enzyme. Furthermore, the overall catalytic efficiency of the mutants increased 1.4- and 1.6-fold as compared to the wild type.

Other examples can be found elsewhere. The second methodology underlines that rational pinpoint modifications in one or more amino acids are made, where these changes are predicted to bring along the envisaged improvement in the targeted enzyme function.

The alterations promoted are performed based on the growing knowledge on the structure and functions of enzyme. Information on this matter mostly comes from bioinformatics, which provides data on amino-acid propensities and on protein sequences. Adequate processing of the data enable the output of generalized rules predicting the effect of mutations on enzyme properties. Also used are molecular potential functions, which, once implemented, enable the prediction of the effect of mutations in enzyme structure.

Computational tools used for enzyme engineering have been recently reviewed. Enzyme engineering through molecular simulations requires structural data from the native enzyme, which can be preferably obtained from crystallography or NMR. Otherwise a model is built based on known enzyme structures with homologous sequences. Computational methods are also welcome in directed evolution, as a tool to better lead the random mutagenesis.

Ultimately this approach is put into practice by producing a site-directed mutant, where selected amino acids are replaced with those suggested from the outcome of modelling. Some relevant examples of this strategy in the area of food and feed processing are given. These mostly aim to improve thermal stability and/or catalytic efficiency and/or to modify the range of pH/temperature where the enzyme is active—goals that were already referred to when examples of enzyme modifications using random mutagenesis were addressed.

The first example underlines the enhancement of the thermostability of the recombinant glucose (xylose) isomerase from Actinoplanes missouriensis and of glucose (xylose) isomerase from Streptomyces diastaticus; of amylases from Bacillus spp. and of glucoamylase from Aspergillus awamori. The mutant isomerase from A. missouriensis displayed an enhanced thermal stability, alongside with improved stability at different pH, as compared with the original enzyme, with no changes in catalytic properties. The double mutant isomerase (G138P, G247D) displayed a 2.5-fold increase in half-life, and additionally a 45% increase in the specific activity, when compared to the wild type.

Such features were ascribed to increased molecular rigidity due to the introduction of a proline in the turn of a random coil. Multiply-mutated amylases obtained by Declerck and coworkers displayed considered enhanced thermal stability. Based on the temperature at which amylase initial activity is reduced by 50% for a 10-minute incubation, this parameter went as high as 106°C, as compared to 83°C for the wild-type strain. Furthermore, the thermal stabilization was not accompanied by a decrease in the catalytic activity.

The work by Lin and coworkers on amylase mutants from Bacillus sp. strain TS-23 highlighted the relevance of E219 for the thermal stability of the enzyme. The mutated glucoamylases engineered by Liu and Wang allowed to establish the role of several intermolecular interactions in thermal stability of these enzymes. Thermostable enzymes were obtained through the introduction of disulfide bonds in highly flexible region in the polypeptide chain of the enzyme, as well as by the introduction of more hydrophobic residues-stabilized α-helices. Data gathered also showed that care had to be taken not to disrupt the hydrogen bond and salt linkage network in the catalytic center as a result of mutagenesis, for this could lead to a decrease in the specific activity and overall catalytic efficiency.

The second example underlines the enhancement of the pH-activity profile and of the thermostability of phytase from A. niger. This was achieved by combining several individual mutations that allowed for mutants that were quite active at pH 3.5. Efficient operation in the stomach of simple-stomached animals where phytate hydrolysis mostly occurs at a pH around 3.5, and the wild type was ineffective, was thus enabled. Furthermore, the hydrolytic activity of the mutants at pH 3.5 exceeded in roughly 1.5-fold that of the parent one at pH 5.5, which was the optimum of the latter. Mutants also retained higher residual activity after incubation within 70 to 100°C, as compared to the wild type.

The work demonstrates that cumulative improvements in pH activity and thermostability through mutation are compatible in this phytase. The third example underlines the modification of the temperature- and pH activity profile of the L-arabinose isomerase from Bacillus stearothermophilus US100. L-Arabinose isomerases catalyze the conversion of L-arabinose to L-ribulose in-vivo, but in-vitro they also isomerize D-galactose into D-tagatose. The latter keto-hexose is being used as a low-calorie bulk sweetener, since its taste and sweetness are roughly equivalent to sucrose, but the caloric value is only 30% of that of sucrose.

Although several thermostable L-arabinose isomerases have been isolated and characterized, most of these display an alkaline pH optimum. For industrial application this presents the same drawbacks of by-product and color formation referred to when the random mutation of glucose isomerases was addressed.

Hence, again arises the need for enzymes able to isomerize L-arabinose in an acidic environment and at relatively low temperature, 60 to 70°C. Operation within the latter temperature range also rules away the use of divalent ions, which stabilize isomerases at high temperatures. Rhimi and coworkers engineered two individual mutants, harbouring each N175H and Q268K mutations.

These led to broader optimal temperature range within 50 to 65°C and to enhanced stability in acidic media, respectively, when compared to the wild type.

An engineered double mutant, harbouring both modifications, displayed optimal activity within a pH range of 6.0 to 7.0 and a temperature range within 50–65°C. Such set of operational conditions matches the targeted goals and again shows that the basis for pH-activity profile and thermostability in L-arabinose isomerase are quite independent and compatible.

Cumulative enhancements in both properties in the same enzyme were thus possible. A similar pattern was also observed in the previous example dedicated to a mutant phytase. The fourth example underlines the modification of the product profile of inulosucrase from Lactobacillus reuteri and from B. subtilis. Inulosucrases are used to synthesize fructooligosaccharides or fructan polymer from sucrose. The transglycosylation catalyzed by the inulosucrase from L. reuteri leads to a wide range of fructooligosaccharides alongside with minor amounts of an inulin polymer.

In order to minimize the dispersion in the products obtained, mutants R423K and W271N were obtained, which allowed the synthesis of a significant amount of polymer and a lower amount of oligosaccharide, without significantly affecting the catalytic activity, when compared with the wild type. The data gathered showed that the "1 subsite in the inulosucrase from L. reuteri has a key role in the determination of the size of the products obtained. Ortiz-Soto and coworkers also showed that the product profile of transfructosylation reactions could be adequately tuned through modification of target residues of an inulosucrase from B. subtilis. These authors established the effect of mutations on the reaction specificity (hydrolysis/ transfructosylation), molecular weight and acceptor specificity.

For example, engineered mutants R360S, Y429N and R433A only synthesized oligosaccharides, whereas the wild type synthesized levan, since the former are more hydrolytic. On the other hand these mutations reduced the affinity for sucrose, and thermal stability, when compared to the wild type.

The fifth example underlines the enhancement of the product profile of cyclodextrin glycosyltransferases (CGTase) from differentgenera. These enzymes promote the production of cyclodextrins, linked oligosaccharides form starch, through an intramolecular transglycosylation reaction. In the process, a starch oligosaccharide is cleaved and cleaved and the resulting reducing-end sugar is transferred to the non-reducing-end sugar of the same chain.

The resulting cyclodextrin may consist of six, seven or eight, which are accordingly termed α, β, or γ-cyclodextrin, respectively. Given their ability to form inclusion complexes with small hydrophobic molecules, they are of interest for both industrial and research applications.

Wild-type CGTases typically produce a mixture of the three cyclodextrins when incubated with starch. The purification of a given cyclodextrin from the reaction mixture requires several additional steps, including selective complexation with organic solvents, which may prove restrictive for cyclodextrin applications involving human consumption.

There is therefore a clear interest in obtaining a mutant CGTase capable of producing a particular type of cyclodextrin in a high rate. Van der Veen and coworkers engineered a double-mutant (Y89D/S146P) of CGTase from Bacillus circulans which displayed a 2-fold increase in the production of α-cyclodextrin and a marked decrease in β-cyclodextrin when compared to the wild type.

From the data gathered, the authors suggested that hydrogen bonds (S146) and hydrophobic interactions (Y89), are likely to play a key role in to the size of cyclodextrin products formed, and that changes in sugar-binding subsites "3 and "7 may result in mutant CGTases with altered product specificity.

Li and coworkers were also able to obtain CGTase mutants from Paenibacillus macerans strain JFB05-01 with increased specificity for α-cyclodextrin, through mutations at subsite "3. In particular, double mutant D372K/Y89R displayed a 1.5-fold increase in the production of β-cyclodextrin, and a significant (roughly 45%) decrease in the production of g-cyclodextrin when compared to the wild-type enzyme.

The two methods are not mutually exclusive and methodologies for engineering of enzymes can assemble both strategies.

Upon identification of the most adequate enzyme, this can be formulated adequately for better process integration. One of the most widely considered approaches for such formulation is enzyme immobilization.

Immobilization

There are several issues that can be lined up to sustain enzyme immobilization. It allows for high-enzyme load with high activity within the bioreactor, hence leading to high-volumetric productivities; it enables the control of the extension of the reaction; downstream process is simplified, since biocatalyst is easily recovered and reused; the product stream is clear from biocatalyst; continuous operation (or batch operation on a drain-and-fill basis) and process automation is possible; and substrate inhibition can be minimized.

Along with this, immobilization prevents denaturation by autolysis or organic solvents, and can bring along thermal, operational and storage stabilization, provided that immobilization is adequately designed. Immobilization has some intrinsic drawbacks, namely, mass transfer limitations, loss of activity during immobilization procedures, particularly due to chemical interaction or steric blocking of the active site; the possibility of enzyme leakage during operation; risk of support deterioration under operational conditions, due to mechanical or chemical stress; and a (still) relative empirical methodology, which may hamper scale up.

Economical issues are furthermore to be taken into consideration when commercial processes are envisaged, although immobilization can prove critical for economic viability if costly enzymes are used.

Still, the cost of the support, immobilization procedure and processing the biocatalyst once exhausted, up- and downstream processing of the bioconversion systems, and sanitation requirements have to be taken into consideration.

In the overall, the enhanced stability allowing for consecutive reuse leads to high specific productivity (massproduct-1 massbiocatalyst-1), which influences biocatalyst-related production costs. A typical example is the output of immobilized glucose isomerase, allowing for 12,000–15,000kg of dry-product high-fructose corn syrup (containing 42% fructose) per kilogram of biocatalyst, throughout the operational lifetime of the biocatalyst.

Increased thermal stability, allowing for routine reactor operation above 60°C minimizes the risks of microbial growth, hence leading to lower risks of microbial growth and to less demanding sanitation requirements, since cleaning needs of the reactor are less frequent. A rule of thumb suggesting that the enzyme costs should be a few percent of the total production costs has been established.

The half-life of the bioreactor is also a critical issue when evaluating the economical feasibility of a bioconversion process, longer half-lives favouring process economics. Examples of commercial bioreactors depict half-lives of several months to years, and the same packing can work throughout some months to years.

Among this group, are immobilized enzyme reactors packed with glucose isomerase for the production of high-fructose corn syrup; lactase for lactose hydrolysis, for the production of whey hydrolysates and for the production of tagatose; aminoacylase for the production of amino acids; isomaltulose synthase for the production of isomaltulose; invertase for the production of inverted sugar syrup; lipases for the interesterification of edible oils, ultimately targeted at the production of trans-free fat, of cocoa butter equivalents, and of modified triacylglycerols; and g-fructofuranosidase for the production of fructooligosaccharides. On the other hand, despite the technical advantages of immobilization, the large-scale liquefaction of starch to dextrins by β-amylases is performed by free enzymes, given the low cost of the enzyme.

Immobilization can be performed by several methods, namely, entrapment/microencapsulation, binding to a solid carrier, and cross-linking of enzyme aggregates, resulting in carrier-free macromolecules. The latter presents an alternative to carrier-bound enzymes, since these introduce a large portion of noncatalytic material.

This can account to about 90% to more than 99% of the total mass of the biocatalysts, resulting in low space-time yields and productivities, and often leads to the loss of more than 50% native activity, which is particularly noticeable at high enzyme loadings.

A typical example of the patterns suggested by data was observed by Abdel-Naby when evaluating the immobilization of β-amylase through different methods. Details on the different methods, as well as some illustrative examples of their applications, are given hereafter.

Entrapment (micro) encapsulation, where the enzyme is contained within a given structure. This can be: a polymer network of an organic polymer or a sol-gel; a membrane device such as a hollow fiber or a microcapsule; or a (reverse) micelle.

Apart from the hollow fiber, the whole process of immobilization is performed in-situ. The polymeric network is formed in the presence of the enzyme, leading to supports that are often referred to as beads or capsules.

Still, the latter term could preferably be used when the core and the boundary layer(s) are made of different materials, namely, alginate and poly-L-lysine. Although direct contact with an adverse environment is prevented, mass transfer limitations may be relevant, enzyme loading is relatively low, and leakage, particularly of smaller enzymes from hydrogels (namely, alginate, gelatin), may occur.

This may be minimized by previously cross-linking the enzyme with multi-functional agent (namely, glutaraldehyde) or by promoting cross-linkage of the matrix after the entrapment. The use of LentiKats, a polyvinyl-alcohol-based support in lens-shaped form, has been used for several applications in carbohydrate processing. Among these are the synthesis of oligosaccharides with dextransucrase, maltodextrin hydrolysis with glucoamylase, lactose hydrolysis with lactase, and production of invert sugar syrup with invertase.

In these processes the biocatalyst could be effectively reused or operated in a continuous manner. Methodologies for large scale production of these supports have been implemented.

Flavourzyme, (a fungal protease/peptidase complex) entrapped in calcium alginate, k-carragenan, gellan, and higher melting-fat fraction of milk fat, was effectively used in cheese ripening, in order to speed up the process, while avoiding the problems associated with the use of free enzyme. These include deficient enzyme distribution, reduced yield and poor-quality cheese, partly ascribed to excessive proteolysis and whey contamination.

The enzyme complex is released in a controlled manner due to pressure applied during cheese curd. Calcium alginate beads were also used to immobilize glucose isomerase and β-amylase for starch hydrolysis to whey.

In the latter work, the authors observed that increasing the concentration of CaCl2 and of sodium alginate to 4% and 3%, respectively, enzyme leakage was minimized (a common drawback of hydrogels) while allowing for high activity and stability.

This effect was also observed in a previous work where alginate-entrapped inulinase was used for sucrose hydrolysis. The stability of an amylase immobilized biocatalyst was further enhanced with the addition of 1% silica gel to the alginate prior to gelation, as reflected by the use of the biocatalyst in 20 cycles of operation, while retaining more than 90% of the initial efficiency.

Several enzymes, namely, chymosin, cyprosin, lactase, Neutrase, trypsin, have also been immobilized in liposomes. In a particularly favored technique immobilization of enzymes in liposomes, known as dehydration-rehydration vesicles (DRVs), small (diameters usually below 50nm) unilamellar vesicles (SUVs) is prepared in distilled water and mixed with an aqueous solution of the enzyme to be encapsulated. The resulting vesicle suspension is then dehydrated under freeze drying or equivalent method.

Upon rehydration, the resulting DRVs are multilamellar and larger (from 200nm to a little above 1000nm) than the original SUVs, and can capture solute molecules. Recent work in this particular application has used lactase as enzyme model and has focused on the optimization and characterization of the liposome-based immobilized system. If liposome-based biocatalysts are used in a process under continuous operation, biocatalyst separation has to be integrated (namely, using an ultra-filtration membrane).

In a different concept, based in batch mode, liposome-encapsulated lactase was incorporated in milk. After ingestion, the vesicles are disrupted in the stomach by the presence of bile salts, allowing in-situ degradation of lactose. Cocktails of enzymes, namely, Flavourzyme, bacterial proteases and Palatase M (a commercial lipase preparation), were immobilized in liposomes and successfully used to speed up cheddar cheese ripening.

Encapsulation in lipid vesicles has been proved a mild method, providing high protection against proteolysis. There is however some lack of consensus on the feasibility of its application on large scale, as well as on the effectiveness of the methodology for controlled release of enzymes. Containment within an ultra-filtration (UF) membrane allows the enzyme to perform in a fully fluid environment; hence, with little loss (if any) of catalytic activity.

However, the membrane still presents a boundary for overall mass transfer of substrate/products and enzyme molecules are prone to interact with the membrane material. This feature is enhanced along with the hydrophobicity of the membrane, hence immobilization in membrane devices may have some adsorptive nature, a feature that will be addressed in (ii).

Besides, regular replacement of the membrane may be required. Enzyme containment by a membrane has been used for the continuous production of galactooligosaccharides from lactose.

The reaction, with up to 80% lactose conversion out of a substrate concentration of 250 gL-1, was carried out in a perfectly mixed reactor and enzyme was recovered in a 10 kDa nominal molecular weight cutoff. The resulting product presented some similarities to the commercially available Vivinal prebiotic. Within the same methodology, a hollow-fiber module was used to contain lactase, in order to carry out lactose hydrolysis in continuous operation. A conversion rate close to 95% in skim milk was observed for an initial substrate concentration close to 40 gL-1.

Binding to a solid carrier, where enzyme-support interaction can be of covalent, ionic, or physical nature. The latter comprehends hydrophobic and van der Waals interactions. These are of weak nature and easily allow for enzyme leakage from the support, namely, after environmental shifts in pH, ionic strength, temperature or even as a result of flow rate or abrasion.

On the other hand, desorption can be turned into an advantage if performed under a controlled manner, since it enables the expedite removal of spent enzyme and its replacement with fresh enzyme. A recent paper by Gopinath andSugunanillustrates the increased trend for leakage when adsorption is compared with covalent binding, using β-amylase as model enzyme.

Curiously, the first reported application of enzyme immobilization was of invertase onto activated charcoal. Recently invertase was immobilized in different types of sawdust, aiming at its application for sucrose hydrolysis.

When wood shavings were used as support, the immobilized invertase retained 90% of the original activity after 20 cycles of 15 minutes, each under consecutive batch operation; and it retained 65% of the original activity after 10 hours of continuous operational regime in a column reactor. Anther example is the immobilization of pectinase in egg shell for the preparation of low-methoxyl pectin. The immobilized biocatalyst could be reused for 32 times at 30°C, and it was used in a fluidized-bed reactor, operated at an optimum flow rate of 5 mL h-1 and 35°C.

Other examples are the surface immobilizations of β-amylase on alumina and in zirconia. Covalent binding is the strongest form of enzyme linking to a solid support. It involves chemically reactive sites of the protein such as amino groups, carboxyl groups, and phenol residues of tyrosine; sulfhydryl groups; or the imidazole group of histidine.

The binding can be carried out by several methods; among them are amide bond formation, alkylation and arylation, or UGI reaction. However, this often brings along loss of activity during the process of immobilization, due to support binding to critical residues for enzyme activity, and steric hindrance, among others.

Examples include the immobilization of β-amylase and of levansucrase on glutaraldehyde-treated chitosan beads, through the glutaraldehyde reaction between the free amino groups of chitosan and the enzyme molecule; the immobilization of pectinase onto Amberlite IRA900 Cl through glutaraldehyde cross-linking; glucoamylase onto dried oxidized bagasse, onto polyglutaraldehyde-activated gelatin, or onto macroporous copolymer of ethylene glycol dimethacrylate and glycidyl methacrylate through the carbohydrate moiety of the enzyme; glucoamylase or invertase immobilized onto montmorillonite K-10 activated with aminopropyltriethoxysilane and glutaraldehyde; and invertase immobilized on nylon-6 microbeads, previously activated with glutaraldehyde and using PEI as spacer; on polyurethane treated with hydrochloric acid, polyethylenimine and glutaraldehyde; on poly(styrene-2-hydroxyethyl methacrylate) microbeads activated with epichlorohydrin; or on poly(hydroxyethyl methacrylate)/glycidyl methacrylate films.

Within this methodology for immobilization, highlight should be given to the introduction of commercial supports (namely, Eupergit, Sepabeads) with a high density of epoxide functional groups aimed at multipoint attachment, typically with the g-amino group of lysine, to confer high rigidity to the enzyme molecule, hence enhancing stabilization.

This methodology has been used for lactase immobilization in magnetic poly (GMA-MMA), formed from monomers of glycidylmethacrylate and ethylmethacrylate, and cross-linked with ethyleneglycol dimethacrylate; for the immobilization of cyclodextrin glycosyltransferases to glyoxylagarose supports for the production of cyclodextrins; or for the immobilization of dextransucrase on Eupergit C. Ionic binding to a carrier involves interaction of negatively or positively charged groups of the carrier with charged amino-acid residues on the enzyme molecules.

Ionic interaction may be favored if enzyme leakage is not an issue, since it allows for support regeneration, unlike immobilization by covalent binding.

Ion-exchanger resins are typical supports for ionic binding; among them are derivatives of cross-linked polysaccharides, namely, carboxymethyl- (CM-) cellulose, CM-Sepharose, diethylaminoethyl- (DEAE-) cellulose, DEAE-Sephadex, quaternary aminoethyl anion exchange- (QAE-) cellulose, QAE-dextran, QAE-Sephadex; derivatives of synthetic polymers, namely, Amberlite, Diaion, Dowex, Duolite; and resins coated with ionic polymers, namely, polyethylenimine (PEI).

Recent examples include the immobilization of invertase in Dowex, in Duolite, in poly(glycidyl methacrylate-co-methyl methacrylate beads grafted with PEI, and in epoxy(amino) Sepabeads; lactase immobilization in PEI-grafted Sepabeads; fructosyltransferase in DEAE-cellulose for the production of fructosyl disaccharides; glucose isomerase in DEAE-cellulose or in Indion 48-R; glucoamylase onto SBA-15 silica and in epoxy(amino) Sepabeads. Ionic binding to Sepabeads-like supports has acknowledged multipoint attachment nature. Enzyme molecules can be modified chemically or genetically modified to enhance immobilization efficiency, an approach followed by Kweon and coworkers, who obtained a cyclodextrin glycosyltransferase fused with 10 lysine residues to improve ionic binding to SP-Sepharose.

Carrier-free macroparticles, where a bifunctional reagent (namely, glutaraldehyde), is used to cross-link enzyme aggregates (CLEAs) or crystals (CLECs), leading to a biocatalyst displaying highly concentrated enzyme activity, high stability and low production costs.

The use of CLEAs is favoured given the lower complexity of the process. This approach is recent, as compared with entrapment and binding to a solid carrier, and there are still relatively few examples of its application to enzymes used in the area of food processing. Among those are following.

First is the immobilization of Pectinex Ultra SP-L, a commercial enzyme preparation containing pectinase, xylanase, and cellulose activities.

The CLEA biocatalyst displayed a slight (30%) in the Vmax, maximal reaction rate/KM ratio, but a significant enhancement in thermal stability (a roughly 10-fold increase in half-life), when the pectinase activity of the immobilized biocatalyst was compared with the free form.

Second is the immobilization of lactase for the hydrolysis of lactose, where, under similar operational conditions as for the free enzyme, the CLEA yielded 78% monosaccharides in 12 h as compared to 3.9% of the free form. Third, CLEAs of glucoamylase, formed by either glutaraldehyde or diimidates, namely, dimethylmalonimidate, dimethylsuccinimidate, and dimethylglutarimidate, led to biocatalysts with improved thermal stability as compared to the free form (over 2-fold increase in half-lives).

Fourth, CLEAs of wild type and two mutant levansucrases were assayed for oligosaccharides/levan and for fructosyl-xyloside synthesis. Although the specific activity of the three free enzymes was 1.25- to 3-fold higher than the corresponding CLEAs, these displayed a 40- to 200-fold higher specific activity than the equivalent Eupergit-C-immobilized enzyme preparations.

Furthermore, all CLEA preparations displayed enhanced thermal stability when compared with the corresponding free enzymes.

Fifth are CLECs of glucose isomerase, aimed at the conversion of glucose into fructose for the production of high fructose corn syrup. When placed in a packed-bed, the resulting enzyme preparation allowed for flow rates that matched or even exceeded those processed by commercially available enzyme preparations (either free, carrier free, or carrier-bound), while achieving the same 45% yield in fructose, under similar operational condition.

Sixth, CLECs of glucose isomerase packed in a column were also used for the concentration/purification of xylitol from dilute or impure solutions. The approach was based on the high specificity of the enzyme crystals towards xylitol, allowing its separation from other sugars, including the natural substrates, xylose and glucose.

Recovery of the adsorbed xylitol was achieved by elution with CaCl2 solutions, with Ca2+ being acknowledged to inactivate glucose isomerase.

Each method for enzyme immobilization has a unique nature. Therefore, despite the potential of immobilization to improve enzyme performance by enhancing activity, stability, or specificity, no specific approach tackles simultaneously these different features.

A careful evaluation and characterization of the methodology addressed is thus required, which can be significantly fastened by high-throughput approaches. Again, the feasibility of its application to reactor configuration and mode of operation has also to be considered in the selection process of the most adequate immobilized biocatalyst for a given bioconversion.

Typical Bioreactors

The most common form of enzymatic reactors for continuous operation is the packed-bed setup, basically a cylindrical column holding a fixed bed of catalyst particles.

These should not have sizes below 0.05mm, in order to keep the pressure drop within reasonable limits. Commercially available carriers such as Eupergit C have particle sizes of roughly 0.1 mm. Commonly operated in down-flow mode, the range of flow rates used must be such as to provide a compromise between reasonable pressure drop, minimal diffusion layer and high conversion yield. Minimization of external mass-transfer resistances with enhanced flow rates can be considered, leading to the fluidized-bed reactor.

This is basically a variation of the packed-bed reactor, but operated in up-flow mode, where the biocatalyst particles are not in close contact which each other; hence, pressure drop is low, and accordingly are pumping costs.

The residence time allowed by the flow rates required for fluidization may however result in low conversion yields. This can be overcome by operating a battery of reactor or by operation in recycle mode. Bioconversions with free enzymes are carried out in stirred tanks.

When on their own, they are restricted to batch mode, but when coupled to a membrane setup with suitable cutoff, they can be integrated in a continuous process, since the enzymes are rejected by the membrane, which acts as an immobilization device, whereas the product (and unconverted substrate) freely permeates. Shear stress induced by stirring creates a hazardous environment for immobilized biocatalysts, particularly when hydrogels are considered, since they are prone to abrasion. In order to overcome this, a basket reactor was developed, but is seldom used, possibly due to mass transfer resistances associated.

Conclusions and Future Perspectives

The integration of enzymes in food and feed processes is a well-established approach, but evidence clearly shows that dedicated research efforts are consistently being made as to make this application of biological agents more effective and/or diversified. These endeavors have been anchoring in innovative approaches for the design of new/ improved biocatalysts, more stable (to temperature and pH), less dependent on metal ions and less susceptible to inhibitory agents and to aggressive environmental conditions, while maintaining the targeted activity or evolving novel activities.

This is of particular relevance for application in the food and feed sector, for it allows enhanced performance under operational conditions that minimize the risk of microbial contamination. It also favours process integration, by allowing the concerted use of enzymes that naturally have diverse requirements for effective application. Such progresses have been made through the ever-continuing developments in molecular biology, the accumulated evolutionary enzyme engineering expertise, the (bio) computational tools, and the implementation of high-throughput methodologies, with high level of parallelization, enabling the efficient and timely screening/characterization of the biocatalysts. Alongside with these strategies, the immobilization of enzymes has also been a key supporting tool for rendering these proteins fit for industrial application, while simultaneously enabling the improvement of their catalytic features. Again, and despite the developments made in this particular field, there is still the lack of a set of unanimously applicable rules for the selection of carrier and method of enzyme immobilization, which furthermore encompass both technical and economic requirements. The latter can be particularly restrictive in the food and feed sector, since most products are of relatively low added value.

Therefore, there is no universal support and method for enzyme immobilization aimed at application in food and feed (let alone the overall range of possible fields of use), and the immobilized biocatalyst fit for a given process and product may be totally unsuitable for another. Given the diversity of enzyme nature and applications this pattern is unlikely to be reversed. Hence, it can be foreseen that efforts will be towards the development of immobilized biocatalyst with suitable chemical, physical, and geometric characteristics, which can be produced under mild condition, that can be used in different reactor configurations and that comply with the economic requirements for large-scale application. All these strategies either isolated or preferably suitably integrated have been put into practice in food and feed, to improve existing processes or to implement new ones, with the latter often combined with the output of new goods, resulting from novel enzymatic activities. Given the recent developments in this field, this trend is foreseen to be further implemented.

Chapter 5

Enzymology for the Food Science

Since the beginning of human history, man has used enzymes indirectly. The fermentation of sugar to ethanol in the preparation of beer and wine, production of vinegar by the oxidation of ethanol, curdling of milk by lactose fermentation are thousands of years old processes where catalytic activities of enzymes are responsible for chemical transformations. Probably the first application of cell free enzyme was in cheese making where rennin obtained from calf stomach was used. The protease rennin which coagulates milk protein, has been used for hundred of years in cheese preparation. The first commercial enzyme was probably reported in Germany in 1914. Use of trypsin, the protease isolated from animals, was shown to improve washing power of detergent over traditional products. Success in the formulation of improved quality of detergent triggered efforts towards the selection of proteases suitable for application in detergents.

Subsequently, a breakthrough in the commercial use of enzyme occurred with the introduction of microbial protease in washing powder at an affordable cost. The first commercial alkaline protease from *Bacillus* sp. was marketed in 1959 and production of enzyme added detergent soon became a big business within a few years. During the period when use of alkaline protease in detergent became popular, use of enzymes in food processing industries also gained momentum in parallel. Fruit clarifying enzyme, called pectinase, was used in fruit juice manufacturing units since 1930. Enzymes hydrolyzing starch into dextrin and glucose largely entered the food industry in 1960 and more or less completely replaced acid process of starch hydrolysis. Starch hydrolyzing enzymes (α-amylase and amyloglucosidase) for the production of glucose soon became the second largest used group of enzymes in industry after detergent protease.

Enzymes may be extracted from any living organism. Sources of commercial enzymes cover a wide range, from microorganisms to plants to animal sources. But for various reasons, microorganisms became the major source of enzymes. In commercial enzyme production, fungi and yeast contribute about 50%, bacteria 25%, animal 8% and plant 4% of the total. Microbes are preferred to plants and animals as they are cheap sources, their enzyme contents are predictable and growth substrates are obtained as standard raw materials. In addition, genetic engineering has opened a new era of advanced enzyme technology. Recombinant DNA technology has made it possible to obtain enzymes present in valued sources, to be synthesized in easy growing microorganisms and also to produce tailor-made enzyme proteins with desired properties as per customers' requirements. Enzymes retaining activity under extreme conditions of temperature, pH and salt concentrations, partially active in organic solvents are all becoming a reality. The prospects of the enzyme industry look very bright with increased market position for existing use, new use of known enzymes and new enzymes having novel industrial applications.

According to a recent release by Business Communication Company Inc. Study RC-147 NA, Norwalk, CT 06855, USA on "Industrial Enzyme Products, Technologies and Applications", the world wide total industrial production of enzymes was of valued at 1.5 billion US dollars in 1997 and it gained an average annual growth rate of 4.0%. Food and animal feed applications of enzymes are constantly dominating the use of industrial enzymes on a worldwide basis.

Large volumes of industrial enzymes are usually not purified and are marketed as concentrated liquid or granulated products with specified enzyme lives. Enzymes used for diagnostic or recombinant technology or in the fine chemical industries need to be highly purified products.

The industrial enzyme market is frequently segmented on the basis of applications, rather than the nature or classes of enzymes as enzymologists classify. Application sectors have been classified in different major sectors with respect to applications of enzymes.

1. Food enzymes
   a. Enzymes for starch processing (amylases for production of glucose)
   b. Sweetener production (glucose isomerase for fructose production)

c Bakery products (xylanase, α -amylase, glucose oxidase for dough conditioning, dough quality, loaf volume)

d. Dairy product (rennin, lactase for milk coagulation and hydrolysis of lactose)

e. Fruit juice (pectinase, cellulase, xylanase for juice clarification, juice extraction)

f. Wine making (glucanase and papain for haze clearance)

2. Enzymes for technological applications
   a. Detergent enzymes (proteinase)
   b. Enzyme for textile (cellulase and laccase for microfibril removal and for improving brightness of cloth)
   c. Enzymes for leather processing (protease, lipase)
   d. Enzymes for paper and pulp processing (xylanase)
   e. Enzymes of analytical use such as:
      i. Uric acid by uricase
      ii. Ethyl alcohol by alcohol dehydrogenase
      iii. Ammonia by L-glutamate dehydrogenase
      iv. Cholesterol by cholesterol oxidase
      v. Glucose by glucose oxidase
      vi. Urea by urease.
3. Enzymes for animal feed
   a. Xylanase for fibre solubility
   b. Phytase for removal of phosphate
4. Enzymes for medical applications
   a. Digestive enzymes: Pancreatic enzymes, mammalian protease (pepsin) plant proteases (Bromelain, papain), fungal amylases.
   b. Enzymes with potential therapeutic applications
      i. Asparaginase and glutaminase hydrolyzing L-aspargine and L-glutamine to aspartic and glutamic acids respectively in the treatment of leukemia.
      ii. Urokinase and streptokinase (plasminogen to plasmin) for dissolving blood clot in heart attack.
      iii. Penicillinase for hydrolyzing penicillin during acute penicillin allergy.
      iv. Hyaluronidase for hydrolyzing hyaluronate in heart attack

v. Collaginase for hydrolyzing collagen in skin cancer

vi. Uricase for oxidizing uric acid in gout

5. Enzymes for clinical and diagnostic applications

a. Enzyme linked immunosorbent Assay (ELISA): Enzymes used are: peroxidase, alkaline phosphatase, S- galactosidase

b. Enzyme multiplied immunoassay technique (EMIT): Enzymes used are lysozyme and malate dehydrogenase

c. Enzymatic analysis of blood constituents: Glucose, uric acid, urea, cholesterol, pyruvate, lactate, triglyceride etc.

Immobilized Enzymes

Enzymes accelerate different chemical reactions with high specificity and are not permanently modified by their participation in reactions. But enzymes are costlier than chemical catalysts, in general, and cost effectiveness of enzyme-based processes could be reached by the repeated use of enzymes. But enzymes remain in solution with products and it is not possible to recover them easily from the reaction mixture.

If they are made insoluble or stationary in active forms, repeated use of an enzyme becomes possible. Immobilization is the process by which an enzyme is made insoluble or stationary with the retention of full or substantial activity. Immobilization is also localization or confinement of enzymes during a process, which permits separation of the enzyme from substrate and product for its repeated use. The use of insoluble form of an enzyme in a process offers a number of advantages such as:

a) Repeated use of the same enzyme as far as practicable,

b) Ability to terminate reaction at any stage by the removal of insoluble enzyme,

c) Recovery of enzyme free product

d) General improvement of enzyme stability.

Techniques used for the immobilization of enzyme activity may be classified as:

a) Physical adsorption of enzyme on inert insoluble carrier

b) Fixing of enzyme on insoluble support by covalent binding

c) Entrapment of enzyme activity in polymerized gel

d) Insolubilization of enzyme by cross-linking with bifunctional reagent.

However, it is understandable that some changes of physical and chemical properties of immobilized enzymes may take place because of the development of new microenvironment around enzyme by the supporting matrix.

The changes are usually expressed to various extents by the altered stability and kinetic parameters of the enzymes. Stability of the enzyme either increases or decreases on immobilization depending on the effect of the microenvironment on stability and denaturation of the enzyme.

Enzymes attached to inorganic matrices were found to be more stable than those attached to organic polymers. Specific activity of an enzyme usually decreases upon immobilization, possibly due to partial denaturation of protein depending on the process of coupling between enzyme and matrix. The presence of electric charge of the matrix affects pH optima of enzymes and the insoluble matrix limits diffusion of high molecular substrates due to steric hindrance.

In the immobilization of an enzyme, it is most important to select the method of attachment, which will not affect or interfere with the substrate-binding site of the enzyme. Considerable knowledge of the active site of the enzyme is essential and any possible interaction with binding site is avoided. Active site of the enzyme is sometimes protected during attachment and freed later.

Use of substrate or competitive inhibitor to protect the active site was found to be useful in some cases. Enzyme as biocatalyst can be immobilized using either purified/semi-purified enzyme (without undesired contaminant activities) or whole cells or sub-cellular components. Most of the enzymes used in industry are microbial extracellular enzymes, which can be isolated more easily from fermented broth as crude enzyme.

Extra cellular enzymes are generally more stable than intracellular enzymes against environmental stress. The cost of enzyme is kept low by the development of fermentation with high enzyme yielding strain. Lengthy and expensive methods of enzyme purification are avoided. Immobilization of whole cell containing the enzyme activity is highly cost effective and the process has some advantages. It preserves the natural environment of the enzyme and loss of activity of immobilized enzyme for each cycle of operation is generally lower than that of free enzyme. Whole cells also provide a number of catalysts present in the cell, if required in a process. But major disadvantages are limitations on the diffusion of substrate to and product from cell and possibility of unwanted side-reactions in presence of other enzyme in whole cell.

Viable and non-viable cells are both immobilized. For single step processes, non-viable cells are used and cells are permeabilized by various physical and chemical treatments. Permeabilization causes diffusion of substrate or product through cell membrane but removes most of the small molecular weight compounds including co-factors from cells. Such non-growing cells are highly useful and economical source of intracellular enzymes for simple bioconversions requiring no regenerated co-factors for activity. A number of techniques have been developed for permeabilization of cells such as treatment of yeast or bacteria with toluene or sonication of cells etc.

There is no best-known method for the immobilization of any specific enzyme. The support, the enzyme, the substrate and technique, are all involved in the development of an effective process. The support material should be non-toxic, low cost, maximum biocatalyst loading capacity and with good flow character and operational durability.

Various Methods Used for Immobilization of Enzymes

In the process of immobilization by binding of enzyme to an inert carrier or cross-linking of enzyme to form an insoluble aggregate, the three dimensional structure of the active site, substrate and catalytic specificities of the enzyme must be kept unchanged during the process of immobilization. Chemical activities, free functional groups of the protein molecule such as α-,β-, γ-, carboxyl groups, free amino groups, thiol, hydroxyl, imidazole groups of amino acids are exploited if they are not involved in formation of catalytic site of the enzyme.

Adsorption on Solid Support

Adsorption of enzyme on solid surface (through Van der waals force or ionic interaction) has advantages because of its simplicity and non-involvement of any chemical reaction, which might affect catalytic activity of the enzyme. But this binding process is a reversible one and desorption of the enzyme becomes significant in presence of high substrate and ionic strength of the medium. Various inorganic matrixes such as alumina, activated carbon, clay, glass, controlled pore glass bead, hydroxy-apatide etc. are used in the immobilization process. Whole cells also could be immobilized using porous brick (yeast), silica (yeast), celite, diatomaceous earth Kieselguhr (mycelial cell), wood chips (yeast) etc. Hydrous metal oxides like those of titanium, zirconium, iron, tin, vanadium are capable of forming active enzyme insoluble complex.

Enzymes may be physically attached to insoluble matrix by ionic interaction. Ion exchange resins, e.g. DEAE-cellulose, DEAE-sephadex,

Carboxymethylcellulose have been used as support media. However this binding is very sensitive to changes of pH and ionic strength of the medium.

Covalent Coupling of Enzyme on Activated Matrix

Covalent binding of the enzyme is more stable than adsorption attachment and enzyme is not leached out easily by changes of pH or at high salt or substrate concentrations. But the process, in general, is more drastic and immobilization is mostly associated with some loss of enzyme activity.

Among the available reactive groups of amino acids, free amino group of protein is predominantly chosen, although coupling through other functional groups is also known.

A few chemical reactions commonly used for the formation of covalent bonds between enzyme and matrices are as follows:

i) The widely used method for immobilization of enzyme on polysaccharide matrix is cynogen bromide activation of matrix. CNBr interacts with –OH groups of the polysaccharide positioned side by side, to produce immidocarbonate derivative. The activated polysaccharide interacts spontaneously with free amino group of amino acid forming covalent bond.

ii) Diazonium derivative of insoluble career (*p*- aminobenzyl cellulose) is used for coupling with phenolic imidazole groups of proteins.

iii) Hydrazide derivative of insoluble matrix is used for coupling

Methyl ester of carboxymethyl cellulose + Hydrazine = hydrazide derivative'! HNO2 '! Azide derivative + NH_2-Enzyme '! Immobilization

iv) Immobilization on synthetic matrix may be exemplified as

Polyaminopolystyrene + $CoCl_2$'!Isocyanate derivative of polystyrene '! NH2- Enzyme (immobilization)

v) Glutaraldehyde is a useful reagent for the formation of covalent linkage between enzyme and matrix. Enzyme could be linked to aminoethyl cellulose or amino alklylated porous glass through this bifunctional reagent,

CHO-CH_2-CH_2-CH_2-CHO. Bis-diazobenzidine also acts as a bifunctional reagent for forming covalent linkage between enzyme and activated matrix.

Entrapment of Enzymes and Microbial Cells

Calcium Alginate

Alginic acid is a co polymer of β-mannouronic acid α-L-Glucuronic acid linked by glycosidic linkages. The polysaccharide is a constituent of marine algae. Alginate solution produces gel in presence of calcium ion. Systems reported such as: Entrapped cells of *Alcaligenes eutrophus* in calcium alginate for the production of H_2 from formate.

Production of isomaltulose from sucrose by *Erwinia rhapontici*

Photosynthesis by immobilized algae.

Carragenan

It is a polysaccharide obtained from red sea algae. It is a linear sulphated polysaccharide containing D-Galactose, 3, 6-anhydro-D-Galactose and their sulphate ester derivative. Among carrageenans, K-carrageenan is insoluble in cold water as potassium salt but is sensitive to Na^+. Some of immobilized systems are:

Production of aspartic acid by entrapped Escherichia *coli.*

Production of ethanol by entrapped Zymomonas *mobilis.*

Agarose

The polysaccharide is obtained form marine plant. It contains alternating 1, 3 linked D-Galactopyranase and 1, 4- linked 3, 6-anhydro-L- Galactopyranose (Linear polymer). Entrapment of *Escherichia coli* in agar beads for the production of hydrogen from formate.

Gelatin

Gelatin is an abundant, inexpensive and safe matrix for immobilization of microbial cells. The protein character, high hydrophilicity and good swelling properties of the matrix are highly favorable for immobilization process. After immobilization, gel is hardened by formaldehyde. A large number of enzyme activities of whole cells were immobilized (e.g. β-glucosidase, urease, invertase and acid phosphatase).

Polyacrylamide Gel

Aqueous solutions of acrylamide, N2 - N2 -ethylene bisacrylamide are polymerized in presence of suitable initiator and accelerator. Enzyme is entrapped within lattice of polyacrylamide gel. ^{11}β-Hydroxylation of cortexolone is done by *Curvularia lunate*, and *Aspergillus niger*.

Cross-linked Pre-polymerized Polyacrylamide

Hydrazide gel partially substituted (acyl-hydrazide) linear polyacrylamide cross linked by glyoxal.

Cephalosporin production by *Steptomyces claruligerus.*

Steroid reduction by *Mycobacterium species.*

Glucose to ethanol by *Sacchromyces cerevisiae.*

Photo Cross-linkable Resin Prepolymers and Urethane

Prepolymer methods use synthetic resin prepolymers, which are photo- cross-linkable resin, polymerize on illumination with UV light. Polyethyleneglycol dimethacrylate, polybutadiene, maleic polybutadiene and polypropylene glycol are used for the purpose. Hydrophobic photo-cross linkable prepolymer (ENT-2000, ENT-4000, ENT-6000, PEGN-1000 / 2000 / 4000) have been used for entrapping sucrase, glucose isomerase, lipase, amino acid deacylase activities.

Water miscible urethane prepolymer (PU) contains isocyanate functional groups at both termini and molecules react with each other in the presence of water. Peroxisome from methanol oxidizing yeast has been successfully immobilized by polyurethane prepolymer.

Micro Capsulation

Synthetic microcapsules are prepared by the use of hydrophilic and hydrophobic monomers. Enzyme is added with hydrophilic monomer in water and mixed with hydrophobic monomer emulsion in organic solvent. Microcapsules of 10-100 ìm diameters are formed with enzyme present in water within. Enzymes are also capsulated in liposomes. Liposomes are lipid membrane with a water droplet inside. They are usually prepared from phosphatidylcholine and cholesterol.

Cross Linking of Enzymes

Same enzyme molecules may be cross-linked by bifunctional reagents such as glutaraldehyde and bis-benzidine. Sufficient cross-linking, occurring at high enzyme concentration makes the enzyme insoluble. However the technique is not much preferred, as it requires large quantities of enzyme and insolubilization is associated with some loss of enzyme activity.

Commercial Processes in Operation

Among the large number of processes reported for the immobilization of commercial enzymes, immobilized enzymes operated successfully in large-scale commercial process are:

a) Isomerisation of glucose for fructose production.
b) Aminoacylase system for amino acid production.
c) Lactase for hydrolysis of whey lactose.

Processes for producing high fructose corn syrup using glucose isomerase use both granules or amorphous or fibrous form of immobilized enzymes with productivities ranging from 1000 - 9000 Kg of 42 % fructose syrup /kg enzyme. The different immobilization procedures used include, DEAE cellulose adsorbed enzyme (*Streptomyces* enzyme), cross-linked lysed cell (*Bacillus coagulans*), gelatin entrapped and glutaraldehyde cross-linked whole cells (*Actinoplanes*), inorganic carrier adsorbed cells (*Streptomyces spp*) etc.

Lactase hydrolyzing lactose into glucose and galactose has a number of applications, particularly in utilizing lactose present in appreciable concentration in whey. In this respect, a large number of immobilization techniques were reported from time to time. Examples include covalent attachment of lactase to glass, collagen, sepharose, entrapment of enzyme in hollow fibre, polyacrylamide gel, and cellulose acetate and in ionizing radiation induced polymers of acrylate and methyacrylate. Lactase immobilized by binding on cellulose sheets was found to be stable for several months. *Aspergillus* lactase immobilized on controlled pore glass or titanium was reported to be available at a very low cost.

Commercial process using aminoacylase system for deacylating acetyl DL– Methionine operated commercially using different immobilized systems. The enzyme attached to DEAE- sephadex by ionic interaction, enzyme covalently bound to iodoacetyl cellulose, enzymes entrapped in polyacrylamide gel were used in different commercial processes.

Production of Glucose from Starch

Starch is synthesized naturally by a variety of plants but some plants produce a high amount of starch such as corn, potato, rice, sorghum, wheat and cassava. Starch contains two types of polysaccharides both containing glucose. An unbranched single chain polymer of 500-2000 or more glucose residues, linked through α-1, 4-glucosidic linkages is called *amylose*. The other fraction called *amylopectin* is branched where branch α- 1, 4-linked glucose chains are linked through α-1, 6- glucosidic linkages to the main chain. The degree of branching in amylopectin is about one per twenty-five residues of unbranched fragment of the main chain.

Starch in general is not soluble in water. It is partially crystalline and quite compact due to intra and inter-molecular hydrogen bonding. When an aqueous suspension of starch is heated at higher temperature, hydrogen bond gradually become weaker and water molecules enter into the starch granules. This process, known as gelatinization, makes starch susceptible to the action of enzyme. Gelatinization is usually accomplished at 90-100° C in water. Addition of alkali to starch suspension and neutralization of alkali also causes gelatinization of starch at room temperature.

Corn Syrup

The knowledge that starch yields a sweet substance when heated with acid was available to chemists two centuries back. Corn emerged as the best source of starch because of its low cost, high availability and long storage life. Acid hydrolysis of cornstarch to produce 42 dextrose equivalent (DE) syrup was commercialized in early 20th century but the product could not compete with sucrose in terms of sweetening power or taste. During 1940-60, the discovery and isolation of amylases from microorganisms led to the development of a number of processes for the production of corn syrup. Various enzymatic hydrolysis protocols were introduced for the production of corn syrup in different food processing industries. Subsequently crystalline dextrose was obtained at 95% yield from cornstarch using amyloglucosidase along with bacterial and fungal α-amylases. The corn hydrolysed products known as "corn sweeteners" are sold on the basis of reducing sugar content or dextrose equivlent (DE). Higher DE value indicates higher degree of hydrolysis. The hydrolyzed product dextrose however remains contaminated with short or medium sized oligosaccharides and their presence lowers the DE of dextrose from 100 to various values. Different types of corn syrup have DE from 20 - 99.5, syrups with DE lower than 20 are usually called as malto-dextrin syrups.

Enzymatic Hydrolysis of Starch

Amylases are the enzymes, which hydrolyse starch to different extents. Enzymes hydrolyzing starch are produced by a wide variety of living beings, including humans. Human saliva and pancreatic secretions contain large amount of α- amylase for starch digestion. However major classes of α- amylases are produced by microbial fermentation. The important enzymes used in starch- saccharification process are α-amylase (EC 3.2.1.1), β- amylase (EC 3.2.1.2), amyloglucoside or glucoamyalse or γ-amylase (EC 3.2.1.3), pullulanase (EC 3.2.1.41) and isoamylase (EC 3.2.1.68). In the hydrolysis of starch, enzymes should specifically hydrolyze both α- 1,4 and α- 1,6 linked glucose molecules in starch.

α- Amylase (α-1,4 glucan-4-glucanohydrolase)

The enzyme is an endo- glucanase, hydrolyzing α-1,4- glucosidic linkages in the interior of the starch molecule. The action of the enzyme is stopped when α- 1,6-glucosidic branch linkages are reached. The hydrolysis of amylose fraction of starch by a α- amylase results in the production of oligosaccharides or dextrins from 1,4-linked glucosidic chain in first phase. Dextrins are subsequently hydrolyzed into maltose and glucose. When amylopectin is the substrate, the hydrolysis products consist of a mixture of unbranched and branched oligosaccharides in which α-1,6-bonds are present The enzymes produced by different groups of bacteria and fungi are classified into liquefying and saccharifying amylases. Liquefying amylases quickly lower the viscosity of gelatinized starch by converting starch into dextrins but slowly hydrolyse dextrin into maltose. Saccharifying amylase, on the other hand, slowly lowers the viscosity of starch by dextrinization but quickly produces maltose and glucose from dextrins. A large number of *Bacillus* species and fungi like *Aspergillus, Penicillium, Mucor* and *Rhizopus* produce saccharifying and liquefying α-amylases. Amylases from *A.oryzae, A. niger, Bacillus amyloliquifaciens, Bacillus licheniformis* are commercially produced.

β - Amylase (α, 1,4- glucan Maltohydrolase)

This enzyme is an exo-enzyme that liberates maltose from linear non-reducing end of the polysaccharide. When β - amylase acts on amylase, it yields maltose quantitatively. But amylopectin is digested by the enzyme to yield maltose and a highly branched dextrin as the enzyme could act on α-1,4-linkages of amylopectin upto 2-3 glucose residues away from the branched 1,6-linkage. Thus, starch which contains both amylase and amylopectin, is hydrolysed by β - amylase into maltose and a highly branched core of amylopectin. Microbial producers of the enzyme are *Bacillus polymyxa, B.megaterium, Streptomyces spps, Pseudomonas spps, Rhizopus japanicus*. Bacterial enzymes are mostly active near pH 7.0 and do not require Ca^{2+} for optimal activity.

Amyloglucoisdase (α-1,4- glucan-glucohydrolases)

The enzyme amyloglucosidase or glucoamylase is an exo-enzyme that liberates α-1,4-linked glucose residues consecutively from non-reducing end of the starch molecule. The enzyme can hydrolyse terminal α-1,6-glucosidic linkages but much slower than α-1,4-linkages. Activity of the enzyme is lowest on maltose and increasingly higher with the increase of oligosaccharide sizes upto 5-6 glucose units.

Fungi are most active producers of amyloglucosidases. Enzymes are obtained from *Apergillus niger, A.oryzae, A.awamori*, and different strains of *Rhizopus*. Amyloglucosidses are frequently produced as isoenzymes by a single strain and all of them are not equally active on cornstarch.

1-6 Glucoside Splitting Enzymes

Enzymes capable of hydrolyzing α-1,6-branching of amylopectin are very important in the saccharification of starch into glucoses. Enzyme hydrolyzing amylopectin directly are pullulanase (EC 3.2.1.41) and isoamyalse (EC 3.2.1.68). Addition of pullulanases during enzymatic hydrolysis of starch by α amylase/ β amylase has a strong synergistic activity on the yield of glucose from starch. Isoamylases are produced by *Bacillus, Serratia*, and *Pseudomonas* spps. Pullulanase is capable of hydrolyzing both pectin and α -glucan microbial polysaccharides including pullulan, but isoamylase acts only on pectin. Pullulanases are produced by bacteria belonging to the genus *Aerobacter, Bacillus, Pseudomonas, Streptococcus*, etc.

The Manufacture of Dextrose from Starch

The process involves the use of thermostable amylase and amyloglucosidase (and pullulanase) for hydrolysis of starch into dextrose monomers at yields close to quantitative. Gelatinized starch is liquefied and dextrinised by thermostable α-amylase. In a typical process this step is carried out in two stages with a holding time of 5-8 minutes at a temperature of 104 °C -107 °C followed by for 90-120 minutes at 94 °C - 97 °C. The gelatinized starch slurry contains suitable does of thermostable α amylase.

The product obtained is soluble dextrin, a mixture of average DE of 10-15. To minimize colour formation due to Maillard reaction, under high temperature the protein content of the slurry is kept low and reaction is carried out at pH 6.0-6.5. The dextrin mixture is hydrolysed by amyloglucosdiase into dextrose. A number of factors determine the final yield of dextrose, such as initial DE of substrate, pH temperature, concentration of solid, amyloglucosidase dose. In a typical process 94-96 % dextrose is obtained by amyloglucosidase treatment of dextrin syrup (30-35 % dry weight) at temperature of 60 °C and pH 4-4.4 for 65-75 hrs.

Enzymatic Hydrolysis of Cellulose into Glucose

The huge plant-biomass, produced every day on earth due to photosynthesis contains cellulose as the major component. Microbial cellulose utilization, on the other hand, is responsible for largest return of the fixed carbon to the biosphere. Cellulose utilizing microorganisms present in soil and in the guts of animals play an important role in completing the carbon cycle on the earth. Cellulose degradation by microorganisms has a major role in anaerobic digestion, composting and for the supply of dietary protein by ruminants.

Cellulose is present in plant biomass in the range of approximately 35 – 50% of plant dry weight. In nature, pure cellulose is obtained from a few sources like cotton, bacterial cellulose etc. In plants, cellulose fibres are embedded in a matrix containing hemicellulose and lignin, which comprise 20 to 35% and 5 – 30% of plant dry weight. The composition of the complex termed as 'lignocellulosics' vary with plant cell type and also with the maturity of the plant. Cellulose is a linear polymer of D-glucopyranose joined together by β-1, 4-glycosidic linkages. Cellulose is synthesized as linear chains, which undergo self-assembly to produce cellulose fibres. Cellulose is initially assembled in elementary fibrils, which are further packed into larger units called micro fibrils, which are finally assembled to cellulose fibres.

Hydrogen bonds and van der waals forces act between adjacent cellulose molecules giving rise to the parallel alignment and crystalline structure of cellulose. The extensive intra- and inter-chain hydrogen bonding produces straight, stable fibres of high tensile strength.

A crystalline structure is a structure where all atoms in the molecule are present in fixed positions relative to each other. Thus crystalline cellulose prevents entry not only of enzyme but also of small molecules due to its extremely tight structure. But this crystalline structure is not uniformly distributed in fibre structure. There are many non-crystalline amorphous regions arising out of twists and other structural variations. These amorphous regions allow penetration of large molecules including cellulolytic enzymes.

Enzymatic hydrolysis of lignocellulose is more complex than that of pure cellulose. Diverse arrangement of cellulose fibres in different plant cells and presence of hemicellulose and lignin makes cellulose hydrolysis difficult.

Cellulose degrading ability is widely distributed amongst different bacterial genera and in many fungal groups. Cellulolytic bacteria include anaerobic Gram positive *Clostridium, Huminococcus*, Gram-negative *Butyrovibrio, Acetivibrio* and aerobic Gram positive *Cellulomonas, Thermobifida.*

Fungi are better known microorganisms decomposing cellulosic substrates; several species of anaerobic fungi and aerobic fungi including *Ascomycete, Basidiomycetes* and *Deuteromycetes* produce cellulolytic enzymes. Cellulolytic enzymes from the genera *Trichoderma, Aspergillus, Penicillum, Fusarium, Geotrichum etc.* have been studied in details. The hydrolysis of cellulose into its monomeric component glucose, involves the participation of a number of enzymes, collectively known as cellulolytic enzymes or cellulases as such.

During the past two decades, extensive biochemical studies have been carried out on cellulase enzymes from aerobic and anaerobic bacteria and fungi. Three major types of enzyme activities identified as components of cellulase enzyme systems are:

i) 1,4- β-D-glucan-4-glucanohydrolase (Endoglucanase, EC 3.2.1.4)

ii) 1,4- β -D-glucan glucanohydrolase (Exo-glucanase, EC 3.2.1.74)
1, 4- β -D-glucan cellobiohydrolase (Exo-glucanase E.C.3.2.1.91)

iii) β -glucosidase or β -glucoside glucohydrolase (EC 3.2.1.21).

In the hydrolytic process, the mechanism of enzyme action was suggested to be as follows.

Endoglucanases cut at internal amorphous sites of cellulose causing release of oligosaccharides of various chain lengths. Exo-glucanases act progressively on cellulose oligosaccharide chains, and liberate either glucose (glucanohydrolase) or cellobiose (cellobiohydrolase) as major products. Finally, β-glucoside hydrolyses soluble cellodextrins and cellobiose into glucose. A general property of most cellulases is their molecular structure, which consisted of a catalytic domain and a carbohydrate binding domain (CBD). CBD facilitates cellulose hydrolysis by bringing the catalytic domain of the enzyme close to the surface of insoluble cellulose substrate.

Cellulase systems are not simply mixture of these three enzymes, but an efficient combination of enzymes, which act, in a coordinated fashion to hydrolyze cellulose into glucose efficiently. Filamentous fungi and actinomycetes have the ability to penetrate cellulosic substrate by the elongation of mycelia and are capable of releasing enzyme on the substrate. These enzyme systems with or without CBM are called 'non-complexed cellulose systems'. Anaerobic bacteria, which cannot penetrate cellulosic substrate, produce 'complexed cellulase system' or cellulosome.

Non-complexed Cellulase System

The cellulase system of aerobic filamentous fungus *Trichoderma reesei* was studied most extensively, followed by that of *Fumicola insolens* and of aerobic actinomycete of genus *Cellulomonas. T. reesei* produces at least five endoglucanases (E.g. I-V), two exo-glucanase (CBH I and CBH II) and two β -glucosidases. Exo-glucanase II and I preferentially attack microcrystalline cellulose from reducing and non-reducing ends of the polysaccharide. They are the major components of cellulase protein of *T.reesei.* The specific role of five individual endo-glucanases is not known clearly, but synergism between *endo-* and *exo-*glucanase has been observed for each enzyme. However, cellobiose the major end product inhibits activities of both *exo-* and *endo-*glucanase. The β-glucosidase produced by *T.reesei* hydrolyses cellobiase and small oligosaccharides to glucose. Both enzymes are present mostly in the cell bound form and little in the supernatant. Cellulase system of *H. insolens* is similar to that of *T. reesei* having two cellobiohydrolases, CBH I and II and five *endo-*glucanases. Cellulase system of cellulomonas is, however, slightly different from that of *T.reesei.* It contains one exo-glucanase and six *endo-*glucanases. However, each enzyme resembles that of *T. reesei* and has CBM.

Complexed Cellulase System

Anaerobic microorganisms produce complexed cellulase system present in 'cellulosome' containing both cellulolytic and non-cellulolytic enzyme activities. Cellulosome is produced as protuberance on the cell wall of anaerobic cellulolytic bacteria like different strains of Clostridia, e.g. *C.thermocellum, C.cellulolyticum, C.cellulovorans, C.josui* etc. and rumen bacteria *Ruminococcus*.

Cellulosome contains a large non-catalytic scaffolding (CipA) protein of 197 KDa size, which is anchored to the cell wall of bacteria. About 22 catalytic proteins are also present in cellulosome. Identified activities are those of nine *endo-*glucanases, four *exo-*glucanases, five xylanases, one chitinase and one lichenase. Cellulosomes are highly glycosylated and are stable protein complexes of 2 to 16 Mda, resistant to protease action.

Acid Hydrolysis of Cellulose for Production of Glucose

The process for the production of glucose by acid hydrolysis of cellulose has many drawbacks. Since hydrolysis of cellulose is to be carried out at high temperature where damage of equipments at high acidity is quick. So cost of hydrolysis equipment is very high also the sugar is spontaneously degraded in acid environment when released.

Additional steps of acid neutralization of hydrolyzed product and purification of product from hydrolysate are also required to be included in the process. A few plants were operated in USA and Europe before World War II.

In the slow acid hydrolysis process, dilute acid was percolated down through packed bed charged with wood chips with continuous removal of sugar solution from the bottom. The reactors were heated with steam at 50 psi. The hydrolysis was continued for 2-4 hour and acid was neutralized with lime. However yield of sugar from cellulose was never higher than 40%. Since hydrolyzed liquor was a dilute sugar solution, recovery of alcohol by fermentation was also costly. Acid hydrolysis for the production of more concentrated solution was also attempted using high-pressure superheated steam. But none of the processes were found to be economic for large-scale production of directly fermentable glucose from cellulose.

Enzymatic Hydrolysis of Cellulose into Glucose

A wide variety of cellulosic substrates are easily available at a low cost, but the main problems associated with low biodegradability of lignocellulosic compounds are:

i) Cellulose is present mostly in crystalline form, which is relatively less accessible to hydrolysis.
ii) Low porosity, which does not allow the cellulolytic enzyme to gain access to the cellulose fibres.
iii) Cellulose micro fibrils are tightly surrounded by lignin, which acts as a cement between cellulose fibres. In addition, hemicellulose (xylan) and other structural polysaccharides form a physical barricade to enzyme action.

Pretreatment of Cellulosic Biomass

Milling of cellulose to 200-mesh increases enzymatic conversion of substrate to sugar by 60-70%. But milling is very energy and capital intensive. The degree of polymerization of cellulose also could be reduced by treatment with alkali, NH_3, or NO, all of which break the fibril hydrogen bond. Treatment of cellulose with ozone or cadoxen also lowers crystallinity of cellulose. High-pressure stem explosion converts the recalcitrant substrate into a hydrolysable substrate for cellulase. But all these pretreatment operations are costly and contribute largely to the cost of production of glucose from cellulosics.

In the commercial production of glucose from cellulose, cost of enzyme, however, is not favourable still today. Highest activity so far, has been reported for enzymes from *Trichoderma reesei* - Rut C-30. Cellulolytic enzymes so far obtained from other microbial sources are also not available at an economic price. Although many pilot plant studies have been carried out, enzymatic hydrolysis of cellulose has not yet been operated on a commercial scale.

Lactase in the Dairy Industry

Lactose is present in milk at about 4.7% (w/v) concentration. Unfortunately, a large fraction of world's adult population shows lactose intolerance and has difficulty in consuming milk and dairy products. Severe tissue dehydrating diarrhea and death are also reported after feeding milk to lactose intolerant children and adults suffering from protein-calorie malnutrition. The availability of low lactose-milk is very important in food-aid programme. In cheese making, the supernatant left after coagulation of milk protein, contains appreciable amount of lactose and the whey cannot be easily disposed off into the environment because of high oxygen demand. On the other hand concentration of whey to syrup causes crystallization of lactose, giving a sandy texture.

Enzyme lactase, a β-galactosidase (EC 3.2.1.23), hydrolyses lactose to glucose and galactose. The hydrolysis of lactose in the milk prevents digestive problems of lactose intolerant population. Lactose hydrolysis has more advantages, it increases sweetness of resultant milk, and it improves scoop and creaminess of ice cream, yoghurt and frozen deserts. The disposal problem of cheese-whey may be managed by the hydrolysis of lactose, when product becomes four times sweeter and whey may be concentrated to microbiologically secured syrup (70%, w/v).

Although a number of microorganisms are known to produce lactase (β-galactosidase), enzymes from *Aspergillus niger, A. oryzae, Kluveromyces fragilis and K .lactis* are supposed to be safe as they have history of safe use for human consumption.

Hydrolysis of lactose is done either by free enzymes in batch process or by immobilized enzyme or by immobilized whole cell. Several commercial immobilized systems have been developed for large-scale operation. In Italy, industrial scale milk processing technology uses fibre-entrapped yeast lactase in a batch process.

Properties of lactase (temperature and pH optima) from different microorganisms determine their applications. Fungal lactases, which work in the acidic range of pH 2.5 – 4.5, are used for acid whey hydrolysis. Yeast and bacterial lactase in neutral pH 6 – 7.5 are suitable for milk and sweet whey hydrolysis. Another important property, determining the application of the enzyme is inhibition of enzyme activity by the product galactose. Enzyme more susceptible to inhibition by galactose is only operative in a dilute solution of whey (at low lactose level) in immobilized column.

Fructose Production

Production of glucose syrup of high dextrose equivalent (DE) from corn became a very successful biotechnology by the use of amyloglucosidase in the saccharification process during 1940 – 50. But sweetness of D-glucose is less (70%) than that of sucrose on weight basis and lower solubility of glucose also posed problems. A commercial syrup containing 71% (w/v) of 97DE glucose has to be kept warm to avoid crystallization while syrup with lower concentrations of glucose is susceptible to microbial contamination. In this respect, fructose consequently received attention as being a sugar with 30% more sweetness than sucrose and two times more soluble than glucose.

Conversion of glucose into fructose requires a single isomerization step. Although chemical isomerization of glucose under alkaline conditions is possible yield is low, reaction is very slow and large numbers of undesired byproducts accumulate in the reaction mixture. Enzymatic conversion of glucose into fructose was known earlier. During glycolysis, the conversion needs participation of three enzymes: hexokinase, phosphohexoisomerase, fructose-6-phosphatase and of ATP. The whole process was commercially unviable, because of unavailability of these enzymes as commercial products and the cost of ATP. The existence of any enzyme capable for isomerizing glucose into fructose was not known before 1950. An enzyme capable of isomerizing D-xylose into D-xylulose was first reported from bacteria. The enzyme initially designated as 'xylose isomerase' was found to catalyze conversion of α-D-glucopyranose into α-D-fructofuranose in presence of cobalt ion. Subsequently, a large number of glucose isomerases were studied onwards from different microorganisms and the name 'xylose isomerase' was replaced by 'glucose isomerase' as enzyme induction did not require xylose to be present as an inducer.

Glucose isomerase (EC 5.3.1.5) catalyses the isomerisation of glucose to fructose. The reaction is reversible and a mixture of fructose and glucose is always obtained by the action of the enzyme on glucose. Most glucose isomerses used today are D-xylose isomerase (Xylose '!xylulose) which are active on D-glucose and also on D- ribose. The enzymes used in different industries usually have high temperature optima. Commercial glucose isomerase producers are *Bacillus coagulans, Streptomyces* and *Actinoplanes* species and *Flavobacterium arborescens.* Some bacteria like *Escherichia intermedia, Aerobacter aerogenes* also isomerise glucose through glucose phosphate isomerase enzyme. The commercial process for the production of glucose from fructose became possible only when enzyme was successfully immobilized. The enzymes are usually intracellular, and usually whole cells or broken cells are used as a source of the enzyme for immobilization.

In the modern process for the production of high fructose syrup, the basic process is enzymatic isomerisation of dextrose by immobilized glucose isomerase. But the process in total, includes many steps, like dextrose production (usually from corn starch), purification of dextrose syrup to make it suitable for feeding to glucose-isomerase immobilized column, isomerisation of dextrose to produce 42% fructose, refining of fructose and fractionation of the syrup for high fructose corn syrup production.

The most important factor in the use of corn dextrose syrup is the purity of the feedstock. The feedstock should have low colour, low ash content and low level of impurities such as metal ions and proteins. The refining process includes use of series of filters to remove traces of particulate matter, decolourizarion and deionization of the filtrate respectively with activated carbon and by cation-anion exchange resins. Magnesium ion is added in the feedstock, which activates glucose isomerase system. Glucose isomersase enzyme is also protected by magnesium from inhibitory action of calcium ion, if present in the feed.

Many immobilized isomerase preparations are available commercially. But cost of the immobilized enzyme is an important factor. The quality of enzyme is judged with respect to longer half – life and higher initial activity.

The conversion of glucose to fructose from 94–96% dextrose attained equilibrium at 60°C, when fructose level reached to 47-48%. But a long residence time of dextrose with enzyme is required to achieve the equilibrium. The conversion level is usually maintained at 42% of fructose with lower residence time. During the process, rate of conversion is proportional to the enzyme activity of the immobilized column. But enzyme activity decreases gradually with time during operation, which either lower fructose concentrate at fixed residence time, or increases residence time for targeted fructose concentration. In operation, a battery of columns is used in place of single one to maintain desired yield at optimum period of time. Another important feature in the process control is the operating temperature and pH. Usually process is conducted at 55°C in pH range of 7.0 – 7.5. Enzyme lives are usually between 3-4 months and columns are generally replaced twice a year.

High Fructose Corn Syrup

Commercial 42% fructose syrup produced frequently does not comply with the quality specifications demanded for the use in soft drinks. Fructose syrup containing more than 55% sucrose was essentially required for high quality soft drinks. The composition of the equilibrium mixture obtained during action of glucose isomerase, ultimately determines the percent of fructose in the reaction product.

Thermodynamics of the system favours fructose production at elevated temperature. The enzyme reactor, which could operate at 95°C, is likely to yield a reaction mixture containing 55% fructose. Water soluble organic solvents also bring similar effects on equilibrium mixture.

But none of the processes were found to be economically viable. Attempts were subsequently made to isolate fructose-rich fraction by various fractionation methods suitable for commercial exploitation.

Chromatography of fructose syrup on Zeolite or calcium salts of cation exchange resin yielded fructose-rich fraction (~90%). The fraction blended with 42% syrup yielded 55% fructose syrup suitable for use in soft drink. Other fractions eluted from ion exchange column, are rich in glucose and contain oligosaccharides. These fractions were treated with glucoamylase to obtain glucose syrup containing 20% solid.

Proteases in Food Processing

Proteases are produced commercially from plants, bacteria and fungi. Enzymes are generally grouped as alkaline, neutral and acid proteases according to their proffered pH range of activities.

Alkaline Protease

Both bacteria and fungi produce alkaline proteases. Various *Bacillus* strains (*B.amyloliquifaciens, B.subtilis, B.megaterium*), Streptomyces strains (*S.fradiae, S.griseus, S.rectus*) and fungi such as *Aspergillus niger*, *Aspergillus flavus* and *Aspergillus oryzae* are potential producers. Subtilisin carlsberg (*Bacillus licheniformis*) and subtilisin BPN (*B. amyloliquifaciens*) are produced in large scales. These enzymes are also called serine proteases, having serine at their active site. Proteases having higher temperature optima, good stability in presence of detergent formulations and working in alkaline range (pH 9-11) are produced commercially in large scale. Alkaline proteases of mammalian origin such as chymotrypsin and trypsin are also serine proteases, having optimum activity near pH 8.0. However these enzymes are structurally different from bacterial serine proteases.

Neutral Protease

Neutral proteases are produced by most of the bacteria and fungi, which produce alkaline proteases. Neutral proteases are relatively less rigid than alkaline proteases, being active in narrow ranges of pH and temperature. Limited uses of these enzymes in the leather and bakery industries have been reported.

Acid Proteases

We have known mammalian gastric enzyme pepsin and rennin of calf stomach for long time. Many fungi also produce acid proteases. Many enzymes are very similar to pepsin and rennin (chymosin). Enzymes produced from *Aspergillus* species, optimally acting in the pH range of 2-4 are used in many pharmaceutical preparations as digestive enzymes. Enzymes are used in industries for breaking gluten of wheat flour or hydrolyzing Soya protein.

Plant Proteases (Bromelain and Papain) in Food Processing

Plant *proteases* such as papain from papaya fruit and bromelain from pineapple have been used in food processing for centuries. These enzymes are thiol proteases having essential cysteine residue in the active site.

Papain present in leaves and unripe fruit of *Carica papaya,* was used to tenderize meats from very ancient times. Papain is commercially produced from the latex of papaya fruit. The latex is obtained by cutting the skin of unripe papaya. Fruits are normally tapped thrice at about 4 – 7 days intervals.

The flow of latex usually ceases 4 – 6 minutes after incision is made. The latex is collected and stored in polythene coated tightly filling boxes, kept under shade. During collection, precaution is taken to avoid contact of latex with heavy metals, such as iron or copper.

All containers including knives, spoons and pots used are either made of plastic or stainless steel. Papain is potentially dangerous, it damages skin on hands under prolonged contact and sometimes causes allergic reaction. The crude papain is obtained by drying the latex. The method of drying largely determines the quality of papain. Crude papain has major use in meat industry for tenderization of meat. Brewing industries in Britain also use papain for chill- proofing beer.

The latex of papaya contains many active proteinases; two major types, papain and chymopapain have been obtained in crystalline forms. Papain has a molecular weight of about 20.5 KDa and isoelectric point at pH 8.8. Papain acts over a wide pH range with optima near pH 6.0. Specificity of peptide bond hydrolysis of papain is similar to those of trypsin or chymotrypsin. It preferentially hydrolyses peptide bonds adjacent to arginine or lysine and tyrosine or phenylalanine. Papain is activated by a variety of substances like glutathione, cysteine, H_2S, HCN etc. and it binds metal ions Cu^{2+} and Hg^{2+}.

Meat of older animals can be tenderized by injecting inactive papain (oxidized disulphide form) into the jugular vein of live animal before slaughter. After slaughtering, reducing conditions developed in the meat reactivate oxidized papain, which tenderizes the meat. But later, this practice was found to be inconvenient. Action of papain in slaughtered animals could not be controlled and quality of meat was found to deteriorate with time.

Bromelain is a glycoprotein of 33 KDa molecular weight present in pineapple juice and stem tissue. It has uses in beer clarification and tenderizing meat similar to papain. However the enzyme is costlier than papain and use of bromelain is comparatively low compared to that of papain.

Protease in Cheese Making

The use of rennet in cheese making is a very old technology known to mankind for several centuries. Rennet (bovine chymosin), obtained from the fourth stomach of unweaned calves has been used for the production of cheese as a milk-clotting agent. Rennet hydrolyses specific peptide linkage between phenylalanine and methionine residues of › -casein protein present in milk. The κ-casein maintains the colloidal character of milk. In the protein molecule, N-terminal hydrophobic domain of ›-casein remains associated with insoluble α- and β-casein while negatively charged C-terminal interacts with water and all these interactions prevent casein micelles from growing large.

Action of rennet hydrolyzing phenylalanine-methionine bond, separates the two domains with the release of hydrophobic para-κ-casein. Termination of the protective action of κ -casein causes coagulation milk to form curd. The coagulated protein is compressed and recovered as cheese. The enzymatic process is temperature dependent and requires presence of calcium ions.

An efficient transformation of milk into gel depends on both the factors. Calf rennet, which contains chymosin and a small amount of pepsin, is a relatively expensive enzyme and has limited supply. Extensive research has been carried out to obtain rennet-like proteases from microorganisms. The major utility of rennet in cheese manufacturing is its proteolytic mode of action and its easy thermal inactivation during ripening of cheese.

Presence of any residual protease was found to cause development of unwanted bitter off-flavours in cheese after ripening. Chymosin is a relatively unstable enzyme. Little activity of the enzyme remains during ripening of cheese. Many microorganisms are known to produce rennet-like proteases. *Rhizomucor pusillus, R. miehei, Aspergillus oryzae, Irpex latis* enzymes are used in cheese manufacture. Milk clotting enzyme from *R. pusillus* is obtained by the solid fermentation of the fungus on wheat bran. Enzymes from *R. michei* and *Endothis parasitica* are obtained from submerged fermentation. Microbial enzymes with lower thermostability are more suitable for cheese making. Loss of protease activity after milk clotting similar to chymosin, is appreciated. This protease is not likely to develop off-smell of cheese during ripening. An interesting example of chemical modification of the protease, successfully lowering its thermo stability was reported.

Enzyme from *M. miehei* was treated with oxidizing agents like par acids or hydrogen peroxide to lower heat stability. The oxidation converts methionine residue of *M. miehei* enzyme to methionine sulphoxide and thermo stability of the enzyme was lowered by 10°C.

The treated enzyme was found to be very similar to chymosin, producing no off-flavour during ripening of cheese. Attempts were also made to clone chymosin into *Escherichia coli* or *Saccharomyces cerevisiae.* The calf pro-chymosin gene was expressed in *E. coli* as insoluble inclusion bodies from which prochymosin was recovered after denaturation. Enzyme produced by *Saccharomyces cerevisiae* containing cloned prochymosin gene mostly remained bound with cellular debris with only 20% in the soluble supernatant. Later chymosin was successfully cloned and expressed in *Kluveromyces lactis* and large-scale production of the enzyme was developed.

Microbial rennets from various microorganisms are marketed under different trade names. Although proteolytic activities of different fungal rennets are considerably different from that of chymosin, but acceptable cheeses are produced by the use of these enzymes.

Other Minor Uses

Heat labile fungal proteases are used in baking processes to hydrolyze gluten (protein fraction) present in flour. Weak-gluten flour is favored for biscuit manufacturing and unavailability of flour of that quality is managed by the addition of protease in dough prepared from high-gluten flour. Proteases are also used for the recovery of protein from parts (bone) of animals and fish. Controlled hydrolysis of Soya protein by proteases is carried out to produce verities of hydrolysates with different new properties.

Enzymes in Leather Industry

One of the oldest industrial uses of enzyme activity is in leather processing. Before raw hide is transformed into leather, it undergoes a series of operations whereby leather making protein collagen present in hides and skins is freed or partially freed from non-collagenous constituents. Hides and skins contain fat as well as globular proteins, *viz.* albumin, globulin, mucoids and fibrous protein such as elastin, keratin and reticulin between collagen fibres.

In industry, raw material is processed through a series of operations including soaking, liming, dehairing, deliming, bating, degreasing and pickling. In pre-tanning operations, skins and hides are subjected to a water soak, which cleans the raw material and loosens the hair. The conventional and most widely used method for dehairing is the treatment of soaked raw material with lime and sodium sulphide. Subsequent deliming is also done to remove adsorbed lime from the hide. Fat present in skins is usually removed with degreasing agents such as soluble lime soap, kerosene, chlorinated hydrocarbons or spirit. In the traditional processing, a large numbers of pollution causing chemicals, lime, sodium sulphide, and solvents are released in effluents, which are toxic and cause environmental pollution.

In addition of chemical treatment, an enzymatic treatment, known, as 'bating' is an essential step to obtain optimum results. During bating, the scud is loosened and many unwanted proteins are removed. Bating makes the grain surface of the finished leather clean, fine and glossy. The traditional practice of bating is an unhygienic process where uncontrolled fermentation of leather is conducted with manures of dog, pigeon or hen as sources of microorganisms. This process gives a desired character to the finished leather. No chemical process has been developed which can substitute the fermentation.

The leather industry has a major problem regarding industrial pollution due to the use of huge amounts of toxic chemicals and biological pollution by addition of unknown microbial load to the environment. Use of enzyme in pre-tanning processes appeared to be a viable alternative technology where pollution problems resulting from tannery effluent could be significantly regulated or restricted.

Enzymes in Pre-tanning

Proteolytic enzyme is the most important enzyme used in pre-tanning process. Enzymes from plant, animals and microbial sources were known to leather industry for a long time. But development of enzyme-based process became bright with the success of production of enzymes at commercial level and availability of enzymatic formulations at cheap rate.

Animal proteases and microbial proteases from bacteria and fungi are used in leather industry. The physicochemical properties of the enzyme such as substrate specificity, temperature and pH stability and pH activity range are very important factors in the application of enzymes in different steps of pre-tanning operations. Microbial enzymes appeared to be ideal source of the proteases. Enzymes with wide pH range of activities could be produced economically from different microbes. Now various commercial enzyme preparations are available for use in leather industry.

Enzymes in Soaking

Proteolytic enzyme combinations (*Aspergillus parasiticus, Aspergillus flavus, Bacillus subtilis, and Aspergillus awamori*) active in natural or alkaline pH ranges are usually used.

Enzymes in Dehairing

Use of enzyme in dehairing is highly desired in leather procesing. It would eliminate the use of sodium sulphide, one of the most toxic chemicals with obnoxious odour. A large number of proteases from *A. flavus, A fumigatus, A. chraceus, A. effuses, Bacillus sp., Streptomyces sp.* have dehairing or hair loosening effects. Potential use of specific protease keratinase from *Streptomyces fradiae*, for dehairing was also indicated. Enzymes from fungal or bacterial sources are allowed to act at pH 10.0 for about 12 – 16 hours, and hair is removed by mechanical means.

Enzymes in Bating

The process of bating is a method for softening hides by treating them in a warm infusion of animal dung. Deliming and proteolytic actions take place simultaneously in bating. A 'bate' usually contains a proteolytic enzyme, a carrier like wood flour and deliming agents like NH_4Cl or $(NH_4)_2SO_4$. Pancreatic enzymes, bacterial and fungal proteases of neutral and alkaline types are used in bating.

Proteases in Detergent Formulation

Protease used in detergent has the largest single market accounting for 25-30% of total enzyme sales. Proteases added to detergents remove protein from clothes soiled with blood, milk and other proteinaceous materials and thus work more efficiently than non-enzyme detergents. The idea of using protease in detergent goes back to the use of pancreatic extract by Roehm in 1913. But economic and technical successes were only achieved in 1960 with the availability of enzymes from bacteria. The basic requirements for detergent proteases are:

i) Availability at low cost from safe microbial source
ii) Capable of working in high alkaline pH of common detergent solution
iii) Sufficiently thermostable with higher temperature optimum
iv) Low or no allergic response for topical use.

Interestingly the enzymes were produced from *Bacillus spps* from the beginning and alkaline proteins from these species represent the lead molecule for the protease group "Subtilisins'. Although subtilisins are classifed as serine proteases by their catalytic mechanism, but they clearly constitute a separate group in terms of their amino acid sequence and three-dimensional structure. Mainly two species, namely *Bacillus licheniformis* and *Bacillus amylolquifaciens* are large scale commercial detergent protease producers in the global market. In addition enzymes from other strains such as *B. clausit*, *B.lentus*, *B. alkalophilus*, *B. halodurans* etc. are used in detergents worldwide.

All these enzymes are fairly non-specific endo proteases, preferentially hydrolyzing the carboxylic side of hydrophobic amino acid residue, but are also capable of hydrolyzing other peptide bonds. They quickly convert protein substrate into small readily soluble peptides. Detergent proteases under different trade names (Alcalase, Savinase, Esperase, Maxatase etc.) are produced commercially. The enzymes are active over a wide pH range up to pH 12 and temperature up to 60°C.

In the fermentation of subtilisins, a large number of variables are manipulated to obtain high yield at low cost. Industrial production processes normally run as large scale using batch fermentations at a high cell density. The composition of fermentation media and downstream processing of the enzyme are not available in details because of company secrecy. Recent reports available for the production of bacillus subtilisins, indicated yields of enzyme in the range 20-28 gm/L of fermentation broth.

Protein engineering with subtilisin has been tried over a long period to obtain new desired quality of subtilisin. Site directed mutagenesis with the replacement of a single amino acid residue adjacent to active serine residue resulted in the production of enzyme stable in presence of hydrogen peroxide. Hydrogen peroxide and peroxy acids are commonly used in detergents as bleaching agents. Extensive replacement of amino acid of subtilisin BPN (*B. licheniformis*) has been patented to obtain improved properties of the protease.

Earlier attempt at the production of detergent protease at commercial scale had problems due to the cases of development of allergic reaction of the workers by the enzyme dust. This was solved by the development of enzyme granules of average 0.5mm diameter.

In this process enzyme in a mixture of inorganic salt and sugar as preservative, is present in an inner core with a protective colloid. This core is coated with polymers made from polyethylene glycol along with a hydrophilic binder. The granulated enzyme was found to be more resistant toward various additives like peroxide, optical brightener, and detergent present in the formulation.

Medical Applications of Enzymes

Therapeutically used enzymes are required in relatively small amounts, but at a very high degree of purity. Many of the enzymes are used topically and orally. A few are injected into blood circulation particularly in the treatment of life threatening disorders like cancer or heart attack.

In general enzymes are foreign proteins, antigenic in nature and elicit an immune response causing severe allergic reaction, particularly when used for longer periods. Antigenic property is modified in several ways, either by covalent modification or entrapment of enzyme within artificial liposomes, synthetic micro spheres and erythrocyte ghosts. The cost of the enzyme is sometimes found to be very high. Urokinase used for dissolving internal blood clots is prepared from human urine and is very costly compared to microbial enzymes. However genetic engineering is an useful technology, which is exploited for the production of the enzymes by microorganisms.

Digestive Enzymes

Aging causes many individuals to suffer from the problem of indigestion. It has been established that after 40 years of age there is a gradual decrease in the production of digestive enzymes. In addition, poor eating habits like less chewing, late eating in the day and habits like excessive consumption of alcohol, refined carbohydrate and fat also cause inadequate production of digestive enzymes. All these problems could be minimized by the intake of digestive enzymes to improve the efficiency of digestion.

Pancreatic enzyme supplement derived from pork pancreas, is an excellent source of digestive enzymes, provides proteases, amylases and lipase. Although the enzyme supplement has a good record of success, its use has some limitations. Pancreatic enzymes are sensitive to low pH values and are destroyed by pepsin secreted from stomach.

A pH sensitive coating has been applied on several pancreatic enzyme preparations, which dissolve only above pH 5.5-6.0. Thus enzyme activities are protected in low pH condition of the stomach. Supplementation of microbial enzymes particularly those produced by different species of *Aspergillus* has many advantages. They provide a large number of digestive enzymes such as proteases, amylase, lipase, lactases, maltase and invertase. They are normally stable over a wide range of pH compared to animal enzymes. They are active and functional for longer distances in the digestive tract. However microbial enzyme should be free from microbial residue and should be a clean and pure product.

Asparaginases

Asparaginase is an important chemotherapeutic agent used in the treatment of acute lymphoblastic leukemia and other lymphoid malignancies. The treatment is based on the fact that tumor cells are deficient in aspartate-ammonia ligase activity, which limits their ability to synthesize the normally non-essential amino acid L-Asparagine.

During growth of tumor cells, L-Asparagine is drawn from body fluids. The enzyme is administered intravenously and it lowers aspargine levels in the blood stream. Asparginase does not affect normal cells as they produce sufficient L-Asparagine for their function. But concentration of extra cellular L-Asparagine is sufficiently lowered by Asparaginase, which restricts growth of tumor cells due to asparagine starvation. The major limitation in the asparginase therapy is the development of clinical hypersensitivity. Chemically modified Asparginase covalently bonded to monomethoxy polyethylene glycol has been found to be useful for the treatment of patients who are sensitive to the native enzyme.

Native Asparaginases, obtained from *Escherichia coli* or *Erwinia chrysanthemi*, are used either alone or in combination therapy for treatment.

Sports Medicine

In sports medicine, therapeutic effects of proteases have been observed. Proteases support, enhance and regulate circulation and the immune system of the individuals, which helps to optimize workouts and muscle maintenance. Treatment of sports injuries with protease has shown good response in the recovery from inflammation and in speedy healing of bruises and swelling. It was reported that proteolytic enzymes obtained from plant sources are more effective than animal enzymes in wound healing of plastic surgery patients.

Streptokinase and Urokinase

In blood plasma, plasmin is an important serine protease, involved in dissolving fibrin clots. It is present in circulation as plasminogen which is activated to plasmin by a number of factors in tissue including tissue plasminogen activator, urokinase plasminogen activator etc.

Streptokinase is an extracelullar enzyme produced by β-hemolytic *Streptococcus*. It can exert fibrinolytic effects through activation of plasminogen. Plasmin in turn attacks fibrin to degrade it into soluble products. Streptokinase is used as an effective and cheap clot dissolving enzyme drug during myocardial infarction and pulmonary embolism. It is given intravenously after the onset of a heart attack to dissolve blood clots in the arteries of the heart wall. Streptokinase is a single chain 47kDa protein secreted by various strains of *Streptococci*.

However, the enzyme obtained from *Streptococci* remains contaminated with a number of unwanted enzymes as present in the extra cellular medium with streptokinase. Streptokinase being a bacterial protein creates immunological problems (allergic reaction) when present in the circulation. Genetically engineered *E.coli* producing a large amount of intracellular streptokinase has been obtained. Attempt has also been made to obtain streptokinase from safer strains of *Bacillus subtilis* by genetic engineering.

Urokinase is also a medically important enzyme used as an anti blood-clotting drug for treatment of heart patients. The enzyme of mammalian origin is isolated from urine. The cost of purification of enzyme is subsequently much higher compared to that of streptokinase. The possibility for the commercial production of the enzyme from human kidney cell line in hollow fibre reactor is under trial.

Glucose Oxidase Electrode

Glucose oxidase (EC 1.1.34) catalyses the oxidation of β-D glucose to D glucono- 1-5 lactone and hydrogen peroxide when molecular oxygen acts as electron acceptor. D-glucono- 1, 5 lactone, however, hydrolyses spontaneously to gluconic acid. Although the enzyme is present in many fungi but that obtained from *Aspergillus niger* has been studied extensively. The enzyme is a dimeric 160KDa protein with one molecule of flavin adenine dinucleotide (FAD) present as cofactor per monomer. FAD is not bound covalently with the apo-enzyme; it can be extracted from enzyme under mild conditions and can be added to apo-enzyme to restore catalytic activity.

The enzyme exhibits a high degree of specificity for β –D-glucose. Michaelis constant of glucose oxidase with glucose is 20mM and with dioxygen is 1.25 mM. 2-deoxy-D-glucose, D-mannose and D-fructose are also oxidized, but at a much slower rates. Glucose oxidase can transfer electron to many artificial electron acceptors other than molecular oxygen. This activity has been used widely for the quantification of serum glucose by colorometric method. In the presence of peroxidase and an electron acceptor dye, oxidation of glucose is estimated by the development colour of reduced dye.

Estimation of glucose is very important in a number of diverse situations. It is important in food industry, in fermentation processes for on line monitoring and most importantly as a clinical indicator of diabetics. Development of a portable easy analytical technique which can give immediate and reliable on-site results, is highly desired for the estimation of glucose.

Biosensor has been developed as the more convenient portable analytical tool for the purposes stated above. A biosensor, in general, is a self contained integrated device capable of providing specific quantitative and semi-quantative analytical informations using a biological recognition signal.. A catalytic biosensor (e.g. Glucose oxidase) is a kinetic device that measures steady state concentration of a transducer-detectable species formed or lost due to a biocatalytic reaction. The development of glucose biosensor became possible, as enthalpy change associated with the oxidation of glucose-by-glucose oxidase is large enough to be detected by a thermistor. It has been shown by the use of radiolabelled dioxygen that oxygen in hydrogen peroxide is derived from dissolved oxygen but not from water molecule. In principle, the of estimation of glucose may be done on with respect to:

i) pH change due to acid production

ii) Oxygen consumption

iii) Hydrogen peroxide production.

In the amperometric estimation based on oxygen consumption, the biosensor is consisted of a platinum electrode where oxygen is reduced with reference to Ag/AgCl reference electrode.

When a potential (-0.6Volt) relative to Ag/AgCl is applied to platinum electrode a current proportional to the concentration of dissolved oxygen is produced. The rate of electrochemical reduction is dependent of oxygen concentration. Depletion of oxygen concentration by the action of glucose oxidase, is detected by the reduction of current between the electrodes. The biosensor is consisted of an electrode compartment containing a platinum cathode and an annular silver electrode, connected through a saturated solution of KCl. The electrode compartment is kept separated from enzyme compartment by a semipermeable thin plastic membrane permeable to oxygen only. This compartment contained immobilized glucose oxidase separated from test solution by a membrane permeable to sustrate and products only. Biosensors of diameters from 1cm to 0.25 mm are commercially available.

The major problems of these biosensors are their dependence on dissolved oxygen concentration. Some mediator (Ferrocinium ion) which transfer electron directly to electrode is also used.

The biosensor used for the determination of unknown glucose concentration is made with glucose oxidase immobilized on the surface of polarographic oxygen electrode. In a typical process glucose oxidase is immobilized on a membrane by the use of bifunctional crosslinking reagent, glutaraldehyde. Biosensor based on glucose coating with glucose oxidase – immobilized gelatin was also developed. Enzyme has been immobilized onto gelatin by cross-linking with Chromium III acetate. The electrode has a response time of 60 seconds with detection limit of 0.25mM glucose. The accuracy was maintained till two months.

Chapter 6

Technological Principles of Pasteurization

Physical and chemical factors which influence pasteurization process are the following:

a. temperature and time;
b. acidity of the products;
c. air remaining in containers.

Pasteurization processes. In pasteurising certain acid juices for example, there are two categories of processes:

Low pasteurization where pasteurization time is in the order of minutes and related to the temperature used; two typical temperature/ time combinations are as following:

63° C to 65° C over 30 minutes or

75° C over 8 to 10 minutes.

Pasteurization temperature and time will vary according to:

- nature of product; initial degree of contamination;
- pasteurized product storage conditions and shelf life required.

In this first category of pasteurization processes it is possible to define three phases:

- heating to a fixed temperature;
- maintaining this temperature over the established time period (= pasteurization time);
- cooling the pasteurized products: natural (slow) or forced cooling.

Rapid, high or flash pasteurization is characterized by a pasteurization time in the order of seconds and temperatures of about 85° to 90° C or more, depending on holding time. Typical temperature/ time combinations are as follows:

88° C (190° F) for 1 minute;

100° C for 12 seconds;

121°C for 2 seconds.

While bacterial destruction is very nearly equivalent in low and in high pasteurization processes, the 121° C/2 seconds treatment give the best quality products in respect of flavour and vitamin retention. Such short holding times, however, require special equipment which is more difficult to design and generally is more expensive than the 63-65 ° C/30 minutes type of processing equipment.

In flash pasteurization the product is heated up rapidly to pasteurization temperature, maintained at this temperature for the required time, then rapidly cooled down to the temperature for filling, which will be performed in aseptic conditions in sterile receptacles. Taking into account the short time and rapid performance of this operation, flash pasteurization can only be achieved in continuous process, using heat exchangers.

Industrial applications of pasteurization process are mainly used as a means of preservation for fruits and vegetable juices and specially for tomato juice.

Thermopenetration. The thermopenetration problem is extremely important, especially in the case of the pasteurization of products packed in glass containers because it is the determining factor for the success of the whole operation.

During pasteurization it is necessary that a sufficient heat quantity is transferred through the receptacle walls; this is in order that the product temperature rises sufficiently to be lethal to micro-organisms throughout the product mass.

The most suitable and practical method to speed up thermopenetration is the movement of receptacles during the pasteurization process. Rapid rotation of receptacles around their axis is an efficient means to accelerate heat transfer, because this has the effect, among others of rapidly mixing the contents.

The critical speed of for this movement is generally about 70 rotations per minute (RPM). This enables a more uniform heating of products, reducing heating time and organoleptic degradation.

Heating may precede or follow packaging. These principles of different temperature time combinations very largely determine the design parameters for heat preservation equipment and commercial practices. The food processor will employ no less than that heat treatment which gives the necessary degree of micro-organism destruction. This is further ensured by periodic inspection by local sanitary authorities or by the importing countries sanitary services. However, the food processor also will want to use the mildest effective heat treatment to ensure highest food quality.

It is convenient to separate heat preservation practices into two broad categories: one involves heating of foods in their final containers, the other employs heat prior to packaging. The latter category includes methods that are inherently less damaging to food quality, where the food can be readily subdivided (such as liquids) for rapid heat exchange. However, these methods then require packaging under aseptic or nearly aseptic conditions to prevent or at least minimise recontamination. On the other hand, heating within the package frequently is less costly and produces quite acceptable quality with the majority of foods and most of our present canned food supply is heated in the package. In practice, therefore, most of the canned food produced locally in developing countries should be heated within the package.

Food Irradiation

Food irradiation is one of the food processing technologies available to the food industry to control organisms that cause food-borne diseases and to reduce food losses due to spoilage and deterioration. Food irradiation technology offers some advantages over conventional processes. Each application should be evaluated on its own merit as to whether irradiation provides a technical and economical solution that is better than traditional processing methods.

Applications

For products where irradiation is permitted, commercial applications depend on a number of factors including the demand for the benefits provided, competitiveness with alternative processes and the willingness of consumers to buy irradiated food products. There are a number of applications of food irradiation. For each application it is important to determine the optimum dosage range required to achieve the desired effect. Too high a dosage can produce undesirable changes in texture, colour and taste of foods.

Shelf-life extension. Irradiation can extend the shelf-life of foods in a number of ways. By reducing the number of spoilage organisms (bacteria, mould, fungi), irradiation can lengthen the shelf life of fruits and vegetables. Since ionising radiation interferes with cell division, it can be used as an alternative to chemicals to inhibit sprouting and thereby extend the shelf life of potatoes, onions and garlic. Exposure of fruits and vegetables to ionising radiation slows their rate of ripening. Strawberries, for example, have been found to be suitable for irradiation. Their shelf-life can be extended three-fold, from 5 to 15 days.

Disinfestation. Ionising radiation can also be used as an alternative to chemical fumigants for disinfestation of grains, spices, fruits and vegetables. Many countries prohibit the importation of products suspected of being contaminated with live insects to protect the importing country's agricultural base. With the banning of certain chemical fumigants, irradiation has the potential to facilitate the international shipment of food products.

Table: *Possible causes of spoilage (real or apparent) in canned goods*

Type of food: Acid and high acid foods (canned fruits)

Condition of can	***Action to be taken to identify cause***
Insufficient vacuum or	Check vacuum and headspace in relation to storage
headspace	temperature and altitude
"Springer" or "flipper"	Cool can to 15°C and check if still domed. Check can for denting, if possible measure headspace volume change brought about by dents, by comparing can volume with volume of a sound can
Hydrogen swell	Check degree of detinning in can especially at the liquid level. Look for scratches or pinholes in lacquer or tin coating. Check if can is still domed on cooling to 15°C.
"Hard" or "soft swell"	Leaker spoilage. Check can for gross seam faults, perforation due to corrosion or damage to seams. Examine contents for signs of spoilage and can interior for detinning at air/ product interface.

Source: FAO/WFP, 1970

Global Developments

Consensus on Wholesomeness

In 1980, an FAD/IAEA/WHO Expert Committee reviewed in detail all the accumulated data on food irradiation from the past 40 years.

The Expert Committee concluded that irradiation to an overall dose of 10 kGy (kilograys) presents no toxicological hazard and introduces no special nutritional or microbiological problems, thus establishing the wholesomeness of irradiated foods up to an overall average absorbed dose of 10 kGy.

Data were insufficient to formulate conclusions on applications of food irradiation above 10 kGy. Data on radiation chemistry, nutritional and microbiological aspects of food treated above 10 kGy is currently being compiled. In 1983, the Codex Alimentarius Commission, an international group that develops global food standards for the FAO and the WHO, incorporated the 1980 Expert Committee's conclusions regarding the wholesomeness of irradiated foods into the Codex General Standard for Irradiated Foods. This proposed international standard was submitted to member countries to accept or to modify according to individual country needs. Currently most countries that allow food irradiation approve its use on a case-by-case basis. The Codex Alimentarius Commission has also adopted a Recommended International Code of Practice for the Operation of Radiation Facilities for the Treatment of Foods. It is intended to serve as a guide for irradiator operators and government regulators.

International Trade

More than 30 countries have given clearances for the use of food irradiation to process some 40 food items and approximately 30 facilities world-wide treat food by irradiation processing. Approvals for additional items are being considered in many countries and many food irradiation facilities are being planned.

It was anticipated in 1988 that by 1990 there could be approximately 50 commercial/demonstration irradiators in 25 countries. Table below shows commercial applications of food irradiation to fruits and vegetables by country.

Table: *International commercial applications of radiation for fruit and vegetables*

Country	*Location (application date)*	*Food Commodity*
Argentina	Buenos Aires (1986)	Spinach
Belgium	Fleurus (1981)	Dehydrated vegetables
Brazil	Sao Paulo (1985)	Dehydrated vegetables
Chile	Santiago (1983)	Dehydrated vegetables onions, potatoes
China	Shanghai (1985	Potatoes
Cuba	Havana (1987)	Potatoes, onions

Contd...

Country	*Location (application date)*	*Food Commodity*
German Dem. Rep	Weideroda (1983)	Onions, garlic
	Spickendorf (1986)	Onions
Japan	Hokkaido (1973)	Potatoes
Korea	Seoul (1985	Garlic powder
Netherlands	Ede (1978)	Dehydrated vegetables
South Africa	Johannesburg (1981)	Dehydrated vegetables
	Tzaneen (1981)	Fruits, onions, potatoes
Thailand	Bangkok (1971)	Onions

Auxiliary Raw Materials

Auxiliary raw materials used in fruit and vegetable processing technologies play a major role in the determination of their physical and chemical characteristics, sensory properties and nutritive value.

Water

Water is one of the essential factors in the activity of the processing centres; according to the final utilisation, water can be classified in three categories:

a. for technological utilisation (when it comes into direct contact with raw materials and enters in the finished product's composition),

b. for steam generators and

c. for receptacle cooling, washing of equipment and general hygiene.

Water for Technological Uses

Water coming into direct contact with the raw materials used for processing (washing, blanching, etc.) or that used as filling liquid of some canned products, must be of drinking water quality in terms of its physico-chemical and microbiological conditions.

More important even than fulfilling drinking water standards, water used for these purposes pests certain specific characteristics related to the technological step or the raw material treated during the processing.

When very hard water is used for blanching vegetables some pecto-calcium and pectomagnesium complexes are formed which starts the hardening of vegetable tissues. This process continues over the pasteurization of the finished product.

When fruit is processed in sugar syrup, the use of hard water for the syrup preparation could induce the formation of a pectin-sugar-acid gel facilitated by the medium pH and presence of calcium salts.

Soft water has negative consequences associated with mineral and hydrosoluble substances and losses during blanching of vegetables.

For some specific products such as peeled tomatoes, green beans and fine texture fruit, the addition of calcium salts (chlorure, sulphate, etc.) is employed to correct for low texture.

The water hardness is an essential factor when used as filling liquid for canned products; ideally, the hardness of the water should be adapted to the raw material species used for canning. Thus a hardness of 3° is good for beans, 5° to 9° for green peas, green beans, and for fruit and vegetables with a tendency to disintegrate should use even harder water.

In the technological process of cucumber and gherkin preservation by natural acidification (lactic fermentation), water hardness has a paramount role. The literature maintains that as far as texture is concerned, the best results are obtained by using very hard water (about 30°); but since high magnesium and salt content has a negative effect on taste, in practice it is recommended that a water hardness of about 10° be used, which gives satisfactory results from both points of view.

Oxygen present in water can act as a corrosion factor in metal receptacles but this negative influence can be eliminated by preliminary boiling.

An important factor is pH. Water for canning must be neutral or slightly basic. Acid water plays a major role in corrosion which is evident both on receptacles and on iron or copper equipment, where changes of product colour will be induced. More dangerous is the attack on lead pipes or to the mix used for can sticking and this can render the product toxic. For these reasons an acid water must be neutralised before use.

Water for Steam Generators

Two main conditions must be fulfilled:

- hardness has to be as low as possible, even zero, because precipitation of calcium salts can lead to the formation of encrustations ("crusts") in pipes and on equipment walls. For this reason, water treatment is practiced in the majority of installations;
- from a bacteriological point of view, the iron-bacteria must be eliminated with biological filters or oxidising substances. This

is necessary in order to avoid iron hydroxide formed during equipment exploitation to deposit on the inner walls of the pipes.

The elimination of iron-bacteria is also of importance for the water used in processing steps.

Water for Receptacle Cooling and General Hygiene

This should be of drinking water standard. Where this is not available in sufficient quantity, the use of industrial water is acceptable but only for cleaning of production rooms/workshops.

Sweeteners

Sugar

Sugar is the conventional name applied to sucrose. Physically there is icing, granulated and lump sugar. In fruit and vegetable processing, sugar is used only in its granulated form; this quality must be in the form of uniform crystals, white, shining and completely soluble in water.

Concentration of various sugar solutions can be rapidly measured by refractometer reading or with areometers graduated in various ways: Brix, Baumé, etc.

Sugar solubility in water is dependent upon temperature; for example, in order to obtain a saturated solution, one must dissolve 2040 g in one litre of water at 20° C and 4870 g at 100° C. Taking into account this temperature related solubility, in practice the majority of sugar solutions are prepared by heating the water. Water should be as soft as possible because the calcium salts can precipitate on boiling.

Corn Syrup (Liquid Glucose)

Corn syrup is obtained industrially by acid or enzymatic starch hydrolysis, using as starting raw materials maize (corn) or potatoes. In fruit processing, mainly in the production of marmalades, it is possible to use corn syrup. The average composition of this corn syrup is of about 32-40% dextrose (glucose), about 40% dextrins and 18-20% moisture. Sweetening power is 50% compared with sucrose.

In a 10%-20% proportion with sucrose, addition of corn syrup has certain advantages:

a. it improves the shine and texture of marmalade;
b. it prevents "sugaring" defect and
c. it reduces the too sweet taste of finished products obtained with sugar alone.

Salt

Salt is used in order to give to the finished products a specifically salty taste and as a preserving substance. From a chemical point of view the term salt means sodium chloride but in practice the product is never in a pure state. The presence of a significant quantity of magnesium chloride increases the hygroscopicity, gives a bitter taste and can induce corrosion of receptacles.

Food Acids

Acetic acid is in use as solutions of various concentrations which are known under the generic name of vinegar. Vinegar can be obtained:

a) from wine, alcohol, cider, beer, etc. by fermentation;
b) by dilution of acetic acid obtained by dry wood distillation or by synthesis.

From a quality point of view, wine vinegar is preferred, as it has a more pleasant flavour. In order to improve taste, other vinegar types are usually flavoured with spices.

In addition to its spicing and flavouring role, vinegar is used and acts as a preservation agent for some vegetables: cucumbers, acidified vegetables, etc.

- Citric acid
- Tartric acid
- Lactic acid.

Pectic Preparations

In fruit processing there many preparations and mixes known as "pectin" are used as liquid or powder extracts.

From a practical point of view, the pectic preparations are classified as:

- strong pectins obtained from apples or citrus fruit peel; this category gives gels rich in sugar;
- weak pectins which gives gels with low proportion of sugar or even without sugar but with the addition of calcium salts.

The fruit industry uses mainly strong pectins. These preparations are characterised, from a commercial point of view, by the capacity of gelification, expressed in degrees. The degree of gelification represents the quantity of sugar in grams able to be transformed in a standard gel (65% sugar and pH=3) by 1 g pectin.

Intensive Sweeteners

"Calorie-reduced" and "low-calorie foods are widely used and are cornering an increasing share of the market. Sweeteners are making an important contribution to the manufacture of sweet foods in these categories. They make it possible to manufacture sweet products without "sweet" being necessarily synonymous with "high-calorie".

Diabetics need to restrict their intake of sugar and various carbohydrates similar to sugar or avoid them altogether. Sweeteners enable diabetics to enjoy sweet tastes without changing their lifestyle. Sweeteners do not contribute to the development of tooth decay; they do not degrade in the mouth to form the acids which are responsible for caries. Thus sweeteners offer consumers a number of advantageous and favourable properties above and beyond merely reducing calories.

Sunett is the trade mark of Hoechst AG for its high intensity sweetener acesulfame K. As an ingredient, it can be used for sweetening all foods produced industrially or at home, or to produce tabletop sweeteners. Like all other sweeteners, Sunett tastes more intensely sweet than sugar; it is about 200 times sweeter than sugar compared with a dilute stock solution.

Synergism. - Sunett is notable for its pronounced synergism with other sweeteners. The synergistic effect leads as quantitative synergism to an intensification of the overall sweetness and as qualitative synergism to an improved taste. The synergism results in a marked intensification of the sweet taste of the blends, which can amount to 30-50% at usual concentrations.

The favourable properties of Sunett, particularly its synergistic behaviour, can be used advantageously in the Sunett-Multi-Sweetener Concept. For high sweetness levels, blends of Sunett with other sweeteners, e.g. aspartame, are particularly favourable in many applications. The synergistic effect of the blend results in a taste that is particularly pleasant and rounded.

Alone or in blends with other sweeteners, Sunett is used mainly where only sweetness is important and no other properties of the foods are to be affected. This applies particularly to beverages (fruit syrups and juices, carbonated beverages, etc.).

In a number of food products, sweet carbohydrates do not only provide the sweet taste; they have other functions to fulfil; for example, they act as bulking or texturing agents and as preservatives by reducing the water activity. In these types of products, Sunett and other sweeteners cannot be used on their own. They must be combined with other substances which perform the required functions. These may be bulking agents or sugar substitutes (for example polyols: sorbitol, mannitol, xylitol). Sunett can be combined with both groups of substances.

Uses. In carbonated soft drinks blends of Sunett with other sweet-tasting substances are often recommended. Beverages formulated using the Sunett Multi-Sweetener Concept, i.e. blends of Sunett with other sweeteners, are usually preferred because of their particularly well balanced and rounded sweetness profile. The taste of these blends is often superiors to single sweeteners.

Sunett is compatible with sugar and other sweet-tasting carbohydrates, both technically and in terms of taste. It can therefore be used in the production of soft drinks with a reduced sugar content. Drinks based on fructose and Sunett are suitable for diabetics. Soft drinks with a reduced sugar content show an excellent taste quality when Sunett or Sunett blends are added to bring them up to the usual level of sweetness.

Fruit nectars and fruit juice drinks differ from most carbonated drinks in that they contain a noticeable amount of sweet carbohydrates provided by the fruit juice. The amount of carbohydrate may vary depending on the type of fruit and amount of juice used.

Amounts of up to 200 mg/l Sunett are often adequate as a single sweetener for the popular types of fruit nectars. With blends of Sunett and aspartame quantities in the order of 100150 mg/l Sunett and 50 mg/l aspartame or 60-70 mg/l of both Sunett and aspartame are sufficient.

Sunett offers such excellent stability that end products, such as drinks, show no reduction in sweetness performance during normal processing methods and storage periods. Sunett can withstand pasteurization, hot filling and aseptic filling without any loss of sweetness.

Jams and marmalades. Sugar contributes a great deal to the texture and stability of conventional jams and marmalades. For the production of sugar-free products either the sugar must be replaced by comparable amounts of sugar substitutes or some of the functions of the sugar must be taken only by other components, such as suitable gelling agents.

Sugar-free jams and marmalades containing sweeteners are more susceptible to microorganisms than sugar-containing products. The risk of spoilage due to yeast fermentation or moulding can be prevented by pasteurization. However, this is only feasible for small jars which will be quickly consumed once they are opened.

In all other cases, it is advisable to add 0.05-0.1% potassium sorbate as a preservative, wherever this is permitted under the relevant food regulations.

For sugar-free jams and marmalades, concentrations in the range of 500-2000 mg Sunett/kg of the finished product are appropriate. It is advisable to add Sunett in the form of an aqueous stock solution towards the end of the boiling process. Care must be taken to ensure that Sunett is evenly dispersed throughout the whole batch. Because of the excellent compatibility of Sunett with sugar alcohols, fruit jams and marmalades using these ingredients offer an outstanding taste.

Fruit preserves. Sunett can be used for the production of sugar-free or sugar-reduced fruit preserves. At the normal pH values for fruit preserves Sunett can be added even before pasteurization, as the sweetness is not impaired under the usual thermal treatment conditions. Sunett also withstands the usual storage periods without loss of sweetness.

NutraSweet (r) is the commercial name of aspartame (APM), a new sweetener from G.D. SEARLE & Co. which can be used in the most foods in order to give the same taste as sugar. NutraSweet is about 180 to 200 times sweeter than sucrose (sugar) and this value depends on pH, temperature and the type of flavour.

NutraSweet can be used as mentioned above in a mix with Sunett or alone in all sugar-free or calorie-reduced fruit jams, marmalades and preserves. Like other sweeteners, NutraSweet does not promote tooth decay.

International regulatory status. Both Sunett and NutraSweet are widely accepted by food laws in the majority of countries. The following are the main fruit products where Sunett is an accepted sweetener:

- low-joule prepared jelly: max. 500 mg/kg;
- canned fruit without added sugar: max. 500 mg/kg;
- beverages including calorie-reduced fruit nectars: max. 600 mg/l;
- calorie-reduced jams;
- calorie-reduced and dietetic fruit compotes;
- canned fruit and vegetables, fruit puree, jams and marmalades;
- juices, nectars and juice based beverages;
- jams, marmalades and related products: max. 300 mg/kg.

Packaging Materials

Requirements and Functions of Food Containers

The following are among the more important general requirements and functions of food packaging materials/containers:

a. they must be non-toxic and compatible with the specific foods;
b. sanitary protection;
c. moisture and fat protection;
d. gas and odour protection;
e. light protection;
f. resistance to impact;
g. transparency;
h. tamperproofness;
i. ease of opening;
j. pouring features;
k. reseal features;
l. ease of disposal;
m. size, shape, weight limitations;
n. appearance, printability;
o. low cost;
p. special features.

Primary and Secondary Containers

The terms primary and secondary containers have been used. Some foods are provided with efficient primary containers by nature, such as nuts, oranges, eggs and the like. In packaging these, we generally need only a secondary outer box, wrap, or drum to hold units together and give gross protection.

Other foods such as milk, dried eggs and fruit concentrates often will be filled into primary containers such as plastic liners which are then packaged within protective cartons or drums. In this case the secondary container provided by the carton or drum greatly minimises the requirements that must be met by the primary container.

Except in special instances, secondary containers are not designed to be highly impervious to water vapour and other gases, especially at zones of sealing, dependence for this being placed upon the primary container.

Since primary containers by definition are those which come in direct contact with the food, we will be far more concerned with them than with secondary containers.

Hermetic Closure

Two conditions of the greatest significance in packaging are hermetic and non-hermetic closure. The term hermetic means a container which is absolutely impermeable to gases and vapours throughout its entirety, including its seams.

Such a container, as long as it remains intact, will automatically be impervious to bacteria, yeasts, moulds, and dirt from dust and other sources since all of these agents are considerably larger than gas or water vapour molecules.

On the other hand, a container which prevents entry of micro-organisms, in many instances will be non-hermetic. A container that is hermetic not only will protect the product from moisture gain or loss, and from oxygen pick-up from the atmosphere, but is essential for strict vacuum and pressure packaging.

The most common hermetic containers are rigid metal cans and glass bottles, although faulty closures can make them non-hermetic. With very rare exceptions flexible packages are not truly hermetic for one or more of the following reasons.

First, the thin flexible films, even when they do not contain minute pinholes, generally are not completely gas and water-vapour impermeable although the rates of gas and water vapour transfer may be exceptionally slow; second, the seals are generally good but imperfect; and third, even where film materials may be gas- and water-vapour-tight, such as certain gages of aluminium foil, flexing of packages and pouches leads to minute pinholes and crease holes.

Hermetic rigid aluminium containers can be readily formed without side seams or bottom end seams. The only seam then to make hermetic is the top end double seam, which may be closed on regular tin can sealing equipment.

Glass containers are hermetic provided the lids are tight. Lids will have inside rings of plastic or cork. Many glass containers are vacuum packed and the tightness of the cover will be augmented by the differential of atmospheric pressure pushing down the cover.

Crimping of the covers, as in the case of pop bottle caps which operate against positive internal pressure, also can make a gas-tight hermetic seal. But bottles fail more often than cans in becoming non-hermetic.

Protection of Food by Packaging Materials

Films and Foils; Plastics

Films and foils have different values for moisture and gas permeability, strength, elasticity, inflammability and resistance to insect penetration and many of these characteristics depend upon the film's thickness.

Important characteristics of the types of films and foils commonly used in food packaging are given in Table below.

For the most part such films are used in the construction of inner containers. Since they are non-rigid, their main functions are to contain the product and protect it from contact with air or water vapour. Their capacity to protect against mechanical damage is limited, particularly when thin films are considered.

These materials can exist in many forms, depending upon such variables as identity and mixture of polymers, degree of polymerisation and molecular weight, spatial polymer orientation, use of plasticisers (softeners) and other chemicals, methods of forming such as casting, extrusion or calendering, etc.

One of the newer classes of plastic materials referred to as copolymers illustrate what can be done with mixtures of the basic units from which plastics are built. The term copolymer refers to a mixture of chemical species in the resin from which films and other forms can be made. The many variations possible make copolymers an important class of plastics to extend the range of useful food packaging applications.

Table: *Properties of packaging films*

Material	*Properties*
Paper	Strength; rigidity; opacity; printability.
Aluminium foil	Negligible permeability to water-vapour, gases and odours; grease proof, opacity and brilliant appearance; dimensional stability; dead folding characteristics.
Cellulose film (coated)	Strength; attractive appearance; low permeability to water vapour (depending on the type of coating used), gases, odours and greases; printability.
Polythene	Durability; heat-sealability; low permeability to water-vapour; good chemical resistance; good low-temperature performance.

Contd...

Material	*Properties*
Rubber hydrochloride	Heat-sealability; low permeability to water vapour, gases, odours and greases; chemical resistance.
Cellulose acetate	Strength; rigidity; glossy appearance; printability; dimensional stability.
Vinylidene chloride copolymer	Low permeability to water vapour, gases, odours and greases; chemical resistance; heat-sealability.
Polyvinyl chloride	Resistance to chemicals, oils and greases; heat-sealability.
Polyethylene terephtalate	Strength; durability; dimensional stability; low permeability to gases, odours and greases.

Source: FAD/WFP, (1970)

Plastic Sheets

- Cellophane paper can be used for packing of dried products, mainly for dried fruit leathers.
- Polyethylene sheets have a variety of uses. They are flexible, transparent and have a perfect resistance to low temperatures and impermeability to water vapour. An important advantage is that these sheets can be easily heat-sealed. Utilisation is in forms of sheets and bags. It is a good packing material for primary protection of dehydrated products. If a good protection is needed to prevent flavour and gas losses, it will be necessary to combine polyethylene with other materials.

Receptacles and Packagings in Plastic Materials

In this class there are three categories:

a. receptacles that can be heat treated: boxes, bottles and bags. Sterilisable bags used up to 120° C can be manufactured from same raw materials as described under plastic sheets and up to 100° C from cellophane. Polyethylene bags could be used to some extent for packing and pasteurization of sauerkraut.

b. receptacles that are not heat treated during processing of fruit and vegetables, also divided in bags and boxes. Bags are the most used type of packing from plastic materials and they are manufactured from polyethylene or cellophane; an important utilisation is for dried/dehydrated fruits and vegetables.

c. special packagings - which can be contacted (Criovac type) by action of heat once the finished product is already inside the pack and the air is evacuated.

Laminates

Various flexible materials such as papers, plastic films, and thin metal foils have different properties with respect to water vapour transmission, oxygen permeability, light transmission, burst strength, pin holes and crease hole sensitivity, etc. and so multi-layers or laminates of these materials which combine the best features of each are used. Commercial laminates containing up to as many as eight different layers are commonly custom designed for a particular product.

Laminations of different materials may be formed by various processes including bonding with a wet adhesive, dry bonding of layers with a thermoplastic adhesive, hot melt laminating where one or both layers exhibit thermoplastic properties, and special extrusion techniques. Such structured plastic films may be complete in themselves or be further bonded to papers or metal foils to produce more complex laminates.

Glass Containers

As far as food packaging is concerned, glass is chemically inert, although the usual problems of corrosion and reactivity of metal closures will of course apply. The principal limitation of glass is its susceptibility to breakage, which may be from internal pressure, impact, or thermal shock, all of which can be greatly minimised by proper matching of the container to its intended use and intelligent handling practices. Here consultation with the manufacturer cannot be over-stressed.

The heavier a jar or bottle for a given volume capacity the less likely it is to break from internal pressure. The heavier jar, however, is more susceptible to both thermal shock and impact breakage. Greater thermal shock breakage of the heavier jar is due to wider temperature differences which cause uneven stress between the outer and inner surfaces of the thicker glass. Greater impact breakage susceptibility of the heavier jar is due to the lower resilience of its thicker wall.

Coatings of various types can markedly reduce each of these types of breakage. These coatings, commonly of special waxes and silicones, lubricate the outside of the glass. As a result, impact breakage is lessened by bottles and jars glancing off one another rather than sustaining direct hits when they are in contact in high speed filling lines.

Surface coating after annealing protects glass surfaces from many of the minute scratches appearing in normal handling after annealing ovens; surface coating also improves the high gloss appearance of glass containers and is said to decrease the noise from glass to glass contact at filling lines.

With regard to thermal shock, it is good practice to minimise temperature differences between the inside and outside of glass containers whenever possible. Some manufacturers will recommend that a temperature difference of 44° C (80° F) between the inside and the outside not be exceeded. This requires slow warming of bottles before use for a hot fill and partial cooling before such containers are placed under refrigeration.

Classification

Glass used for receptacles in fruit and vegetable processing is a carefully controlled mixture of sand, soda ash, limestone and other materials made molten by heating to about 1500° C (2800° F).

Main classes of glass receptacles are:

a. jars which are resistant to heat treatments,
b. jars, glasses, etc. for products not submitted to heat treatment (marmalades, acidified vegetables, etc.);
c. glass bottles for pasteurized products (tomato juice, fruit juices, etc.) or not pasteurized (syrups) and
d. receptacles with higher capacity (flasks, etc.)

Jars for Sterilised/Pasteurized Canned Products

These receptacles may replace metal cans, taking into considering both the advantages and disadvantages they present. Advantages are: they do not react to food content; they are transparent and can be manufactured in various shapes; they use cheap raw materials and are reusable. Disadvantages: heavier than metal can of same capacity; fragile; lower thermal conductance and a limited resistance to thermal shocks.

Receptacles in this category must assure a perfect hermeticity after their pasteurization/sterilisation and cooling and this has to be achieved by the use of metallic (or glass) caps and specific materials for tightness. Taking into account the receptacles' closure method, there are two categories of receptacles:

a. glass jars with mechanical closure;
b. glass jars with pneumatic closure.

Jars for Products without Heat Treatment

For marmalades, jellies and jams glass jars with non hermetic closures using metal, glass or rigid plastic caps are used; however for these products the receptacles mentioned above may also be used.

The use of jars with pneumatic closure presents the advantage that some products (e.g. marmalades, jams) can be filled hot and therefore sterile in receptacles. Pneumatic closing generally protects against negative air action which is in this case eliminated from receptacles.

Glass Bottles

These receptacles are widely used both for

a) finished products which need pasteurization (ea. tomato juice, Knit juices, etc.) and for

b) those which are preserved as such (ea. fruit syrups).

Glass bottles in category a) are closed hermetically with metallic caps, provided with special materials for tightness. For glass bottles in category b) various corks, and aluminium caps with tightness materials may be used.

Glass Receptacles with High Capacity

In this category there are glass flasks with 3 and 10 litre capacity which can be hermetically closed by a SKO caps system and are resistant to product pasteurization (ea. tomato juice). As bigger receptacles it is possible to use glass demijohns with usual capacity of 25 and 50 litre; these receptacles are used for preservation of fruit juices by warm process. Closing is performed with flexible rubber hoods.

Paper Packaging

As primary containers few paper products are not treated, coated or laminated to improve their protective properties.

Paper from wood pulp and reprocessed waste paper will be bleached and coated or impregnated with such materials as waxes, resins, lacquers, plastics, and laminations of aluminium to improve water vapour and gas impermeability, flexibility, tear resistance, burst strength, wet strength, grease resistance, sealability, appearance, printability, etc.

Paper Sheets

- Kraft paper is the brown unbleached heavy duty paper commonly used for bags and for wrapping; it is seldom used as a primary container;

- parchment paper: acid treatment of paper pulp modifies the cellulose and gives water and oil resistance and considerable wet strength to this type of packaging material;
- glassine-type papers are characterised by long wood pulp fibres which impart increased physical strength;
- paper with plastic material sheets.

Receptacles from Paper or Cardboard

(paper = 8 to 150 g/m²; cardboard = 150 to 450 g/m²).

"Tin can"/tinplate

The "tin can" is a container made of tinplate.

Tinplate, a rigid and impervious material, consists of a thin sheet of low carbon steel coated on both sides with a very thin layer of tin. It can be produced by dipping sheets of mild steel in molten tin (hot-dipped tinplate) or by the electro-deposition of tin on the steel sheet (electrolytic tinplate). With the latter process it is possible to produce tinplate with a heavier coating of tin on one surface than the other (differentially coated).

Tin is not completely resistant to corrosion but its rate of reaction with many food materials is considerably slower than that of steel. The effectiveness of a tin coating depends on:

a. its thickness which may vary from about 0.5 to 2.0 μm (20 to 80 x 10(-6) in.);
b. the uniformity of this thickness;
c. the method of applying the tin which today primarily involves electrolytic plating;
d. the composition of the underlying steel base plate;
e. the type of food, and
f. other factors.

Some canned vegetables including tomato products actually owe their characteristics flavours to a small amount of dissolved tin, without which these products would have an unfamiliar taste. On the other hand, where tin reacts unfavourably with a particular food the tin itself may be lacquer coated. The thickness of tinplate sheets may vary from 0.14 mm to 0.49 mm and is determined by weighing a sheet of known area and calculating the average thickness.

Tinplate sheets may be lacquered after fabrication to provide an internal or external coating to protect the metal surface from corrosion by the atmosphere or through reaction with the can contents.

They may also be printed by lithography to provide suitable instructions or information on containers fabricated from tinplate sheets (otherwise paper labels can be attached to the outer tinplate surface).

Under normal conditions the presence of the tin coating provides a considerable degree of electrochemical protection against corrosion, despite the fact that in both types of tinplate the tin coating is discontinuous and minute areas of steel base plate are exposed. With prolonged exposure to humid conditions, however, corrosion may become a serious problem.

The coatings not only protect the metal from corrosion by food constituents but also protect the foods from metal contamination, which can produce a host colour and flavour reactions depending upon the specific food. Particularly common are dark coloured sulphides of iron and tin produced in low acid foods that liberate sulphur compounds when heat processed, and bleaching of red plant pigments in contact with unprotected steel, tin, and aluminium.

Table: *General types of can coatings*

Coating	*Typical uses*	*Type*
Fruit enamel	Dark coloured berries, cherries and other fruits requiring protection from metallic salts	Oleoresinous
C-enamel	Corn, peas and other sulphur-bearing products	Oleoresinous w. suspended zinc oxide
Citrus enamel	Citrus products and concentrates	Modified oleoresinous
Beverage can enamel	Vegetable juices; red fruit juices; highly corrosive fruits; non-carbonated beverages	Two-coated w. resinous base coat and vinyl top coat

Fruit Specific Preservation Technologies

Fruit Quality

Fruit quality goes back to tree stock, growing practices and weather conditions. Closer to the shipper and processor, however, are the degrees of maturity and ripeness when picked and the method of picking or harvesting.

There is a distinction between maturity and ripeness of a fruit. Maturity is the condition when the fruit is ready to eat or if picked will become ready to eat after further ripening. Ripeness is that optimum condition when colour, flavour and texture have developed to their peak. Some fruit is picked when it are mature but not yet ripe. This is especially true of very soft fruit like cherries and peaches, which when fully ripe are so soft as to be damaged by the act of picking itself.

Further, since many types of fruit continue to ripen off the tree, unless they were to be processed quickly, some would become overripe before they could be utilised if picked at peak ripeness.

From a technological point of view, fruit characterisation by species and varieties is performed on the basis of physical as well chemical properties: shape, size, texture, flavour, colour/pigmentation, dry matter content (soluble solids content), pectic substances, acidity, vitamins, etc. These properties are directly correlated with fruit utilisation. Table below shows which of the above mentioned properties have a major impact on the finished products obtained by fruit processing.

Table: *Optimal use of fruits as a function of their properties*

Processed	***Organoleptic (Sensory) Properties***				***Chemical Composition***		
Finished Products	***Shape***	***Texture***	***Flavour***	***Taste***	***Acidity***	***Sugars***	***Pectic Sust.***
Dried Fruits	++	++		++		++	
Fruit Juices			++	++	++		
Marmalade			++	++			++
Jams	++	++	++	++			
Jellies	++	++	++	++			++
Fruit Paste				++		++	

When to Pick

The proper time to pick fruit depends upon several factors; these include variety, location, weather, ease of removal from the tree (which change with time), and purpose to which the fruit will be put.

For example, oranges change with respect to both sugar and acid as they ripen on the tree; sugar increases and acid decreases. The ratio of sugar to acid determines the taste and acceptability of the fruit and the juice. For this reasons, in some countries there are laws that prohibit picking until a certain sugar-acid ratio has been reached.

In the case of much fruit to be canned, on the other hand, fruit is picked before it is fully ripe for eating since canning will further soften the fruit.

Quality Measurements

Many quality measurements can be made before a fruit crop is picked in order to determine if proper maturity or degree of ripeness has developed. Colour may be measured with instruments or by comparing the colour of fruit on the tree with standard picture charts.

Texture may be measured by compression by hand or by simple type of plungers.

As fruit mature on the tree its concentration of juice solids, which are mostly sugars, changes. The concentration of soluble solids in the juice can be estimated with a refractometer or a hydrometer. The refractometer measures the ability of a solution to bend or refract a light beam which is proportional to the solution's concentration. A hydrometer is a weighted spindle with a graduated neck which floats in the juice at a height related to the juice density.

The acid content of fruit changes with maturity and affects flavour. Acid concentration can be measured by a simple chemical titration on the fruit juice. But for many fruits the tartness and flavour are really affected by the ratio of sugar to acid.

Percentage of soluble solids, which are largely sugars, is generally expressed in degrees Brix, which relates specific gravity of a solution to an equivalent concentration of pure sucrose.

In describing the taste of tartness of several fruits and fruit juices, the term "sugar to acid ratio" or "Brix to acid ratio" are commonly used. The higher the Brix the greater the sugar concentration in the juice; the higher the "Brix to acid ratio" the sweeter and lees tart is the juice.

Harvesting and Preprocessing

The above and other measurements, plus experience, indicate when fruit is ready for harvesting and subsequent processing.

A large amount of the harvesting of most fruit crops is still done by hand; this labour may represent about half of the cost of growing the fruit. Therefore, mechanical harvesting is currently one of the most active fields of research for the agricultural engineer, but also requires geneticists to breed fruit of nearly equal size, that matures uniformly and that is resistant to mechanical damage.

A correct manual harvesting includes some simple but essential rules:

- the fruit should be picked by hand and placed carefully in the harvesting basket; all future handling has to be performed carefully in order to avoid any mechanical damage;

- the harvesting basket and the hands of the harvester should be clean;
- the fruit should be picked when it is ready to be able to be processed into a quality product depending on the treatment which it will undergo.

It is worth emphasising the fact that the proximity of the processing centre to the source of supply for fresh raw materials presents major advantages; some are as follows:

- possibility to pick at the best suitable moment;
- reduction of losses by handling/transportation;
- minimises raw material transport costs;
- possibility to use simpler/cheaper receptacles for raw material transport.

Once it has left the tree, the organoleptic properties, nutritional value, safety and aesthetic appeal of the fruit deteriorates in varying degrees. The major causes of deterioration include the following:

a. growth and activity of micro-organisms;
b. activities of the natural food enzymes;
c. insects, parasites and rodents;
d. temperature, both heat and cold;
e. moisture and dryness;
f. air and in particular oxygen;
g. light and
h. time.

The rapidity with which foods spoil if proper measures are not taken is indicated in Table below.

Table: *Useful storage life of some food products*

Food Products	*Generalized Storage Life (Days) at 21°C (70°F)*
Animal Flesh, Fish, Poultry	1-2
Dried, salted, smoked meat and fish	360 and more
Fruits	1-7
Dried Fruits	360 and more
Leafy Vegetables	1-2
Root Crops	7-20

Source: Desrosier and Desrosier (1977)

Reception - Quality and Quantity

Fruit reception at the processing centre is performed mainly for following purposes:

- checking of sanitary and freshness status;
- control of varieties and fruit wholeness;
- evaluation maturity degree;
- collection of data about quantities received in connection to the source of supply: outside growers/farmers, own farm.

Variety control is needed in order to identify that the fruit belongs to an accepted variety as not all are suitable for different technological processes.

Fruit maturity degree is significant as industrial maturity is required for some processing/preservation methods while for others there is the need for an edible maturity when the fruit has full taste and flavour. Special attention is given to size, appearance and uniformity of fruit to be processed, mainly in the form of fruit preserved with sugar using whole/half fruits ("with fruit pieces").

Some laboratory control is also needed, even if it not easy to precisely establish the technological qualities of fruit because of the absence of enough reliable rapid analytical methods able to show eventual deterioration. The only reliable method for evaluating the quality is the combination of data obtained through organoleptic/taste controls and by simple analytical checks which are possible to perform in a small laboratory: percentage of soluble solids by refractometer, consistency/texture measured with simple penetrometers, etc.

Temporary Storage before Processing

This step has to be as short as possible in order to avoid flavour losses, texture modification, weight losses and other deterioration that can take place over this period.

Some basic rules for this step are as follows:

- keep products in the shade, without any possible direct contact with sunlight;
- avoid dust as much as possible;
- avoid excessive heat;
- avoid any possible contamination;
- store in a place protected from possible attack by rodents, insects, etc.

Cold storage is always highly preferred to ambient temperature. For this reason a very good manufacturing practice is to use a cool room for each processing centre; this is very useful for small and medium processing units as well.

Washing

Harvested fruit is washed to remove soil, micro-organisms and pesticide residues.

Fruit washing is a mandatory processing step; it would be wise to eliminate spoiled fruit before washing in order to avoid the pollution of washing tools and/or equipment and the contamination of fruit during washing.

Washing efficiency can me gauged by the total number of micro-organisms present on fruit surface before and after washing - best result are when there is a six fold reduction. The water from the final wash should be free from moulds and yeast; a small quantity of bacteria is acceptable.

Fruit washing can be carried out by immersion, by spray/showers or by combination of these two processes which is generally the best solution: pre-washing and washing. Some usual practices in fruit washing are:

- addition of detergents or 1.5% HCl solution in washing water to remove traces of insect-fungicides;
- use of warm water (about 50°C) in the pre-washing phase;
- higher water pressure in spray/shower washers.

Washing must be done before the fruit is cut in order to avoid losing high nutritive value soluble substances (vitamins, minerals, sugars, etc.).

Sorting

Fruit sorting covers two main separate processing operations:

a. removal of damaged fruit and any foreign bodies (which might have been left behind after washing);

b. qualitative sorting based on organoleptic criteria and maturity stage.

Mechanical sorting for size is usually not done at the preliminary stage. The most important initial sorting is for variety and maturity.

However, for some fruit and in special processing technologies it is advisable to proceed to a manual dimensional sorting (grading).

Trimming and Peeling (Skin Removal)

This processing step aims at removing the parts of the fruit which are either not edible or difficult to digest especially the skin.

Up to now the industrial peeling of fruit and vegetables was performed by three procedures:

a. mechanically;

b. by using water steam;

c. chemically; this method consists in treating fruit and vegetables by dipping them in a caustic soda solution at a temperature of 90 to 100° C; the concentration of this solution as well as the dipping or immersion time varying according to each specific case.

Cutting

This step is performed according to the specific requirements of the fruit processing technology.

Heat Blanching

Fruit is not usually heat blanched because of the damage from the heat and the associated sogginess and juice loss after thawing. Instead, chemicals are commonly used without heat to inactivate the oxidative enzymes or to act as antioxidants and they are combined with other treatments.

Ascorbic/Citric Acid Dip

Ascorbic acid or vitamin C minimises fruit oxidation primarily by acting as an antioxidant and itself becoming oxidised in preference to catechol-tannin compounds. Ascorbic acid is frequently used by being dissolved in water, sugar syrup or in citric acid solutions.

It has been found that increased acidity also helps retard oxidative colour changes and so ascorbic acid plus citric acid may be used together. Citric acid further reacts with (chelates) metal ions thus removing these catalysts of oxidation from the system.

Sulphur Dioxide Treatment

Sulphur dioxide may function in several ways:

- sulphur dioxide is an enzyme poison against common oxidising enzymes;
- it also has antioxidant properties; i.e., it is an oxygen acceptor (as is ascorbic acid);

- further SO2 minimises non enzymatic Maillard type browning by reacting with aldehyde groups of sugars so that they are no longer free to combine with amino acids;
- sulphur dioxide also interferes with microbial growth.

In many fruit processing pre-treatments two factors must be considered:

a. sulphur dioxide must be given time to penetrate the fruit tissues;
b. SO_2 must not be used in excess because it has a characteristic unpleasant taste and odour, and international food laws limit the SO_2 content of fruit products, especially of those which are consumer oriented (e.g. except semi-processed products oriented to further industrial utilisation).

Commonly a 0.25 % solution (except for semi-processed fruit products which are industry oriented and use a 6% solution) of SO_2 or its SO_2 equivalent in the form of solutions of sodium sulphite, sodium bisulphite or sodium/potassium metabisulphite are used. Fruit slices are dipped in the solution for about two to three minutes and then removed so as not to absorb too much SO_2. Then the slices are allowed to stand for about one to two hours so that the SO_2 may penetrate throughout the tissues before processing. Sulphur dioxide is also used in fruit juice production to minimise oxidative changes where relatively low heat treatment is employed so as not to damage delicate juice flavour. Dry sulphuring is the technological step where fruit is exposed to fumes of SO_2 from burning sulphur or from compressed gas cylinders; this treatment could be used in the preparation of fruits (and some vegetables) prior to drying/dehydration.

Sugar Syrup

Sugar syrup addition is one of the oldest methods of minimising oxidation. It was used long before the causative reactions were understood and remains today a common practice for this purpose. Sugar syrup minimises oxidation by coating the fruit and thereby preventing contact with atmospheric oxygen. Sugar syrup also offers some protection against loss of volatile fruit esters and it contributes sweet taste to otherwise tart fruits. It is common today to dissolve ascorbic acid and citric acid in the sugar syrup for added effect or to include sugar syrup after an SO_2 treatment.

Fresh Fruit Storage

Some fruit species and specially apples and pears can be stored in fresh state during cold season in some countries' climatic conditions..

Fruit for fresh storage have to be autumn or winter varieties and be harvested before they are fully mature. This fruit also has to be sound and without any bruising; control and sorting by quality are mandatory operations. Sorting has to be carried out according to size and weight and also by appearance; fruit which is not up to standard for storage will be used for semi-processed product manufacturing which will be submitted further to industrial processing.

Harvested fruit has to be transported as soon as possible to storage areas. Leaving fruit in bulk in order to generate transpiration is a bad practice as this reduces storage time and accelerates maturation processes during storage.

In order to store large quantities of fruit, silos have to be built.

Technology of Semi-processed Fruit Products

The semi-processed fruit products are manufactured in order to be delivered to industry processing centres (in the fruit producing country itself or in importing countries) where they will be further manufactured in consumer oriented finished products: jams, jellies, syrups, fruits in syrup, etc.

In the practice of semi-processed fruit products and for the purpose of this document the following categories are defined:

a. fruit "pulps": semi-processed products, not refined, obtained by mechanical treatment (or, less often, by thermal treatment) of fruit followed by their preservation. Either whole fruit, halves or big pieces are used which enables easy identification of the species. "Pulps" can be classified in boiled or non boiled (raw).
b. fruit "purées-marks": semi-processed products obtained by thermal and mechanical treatment or, very rare, raw and then refined, operations by which all nonedible parts (cores, peels, etc.) are removed. "Purées-marks" are classified in boiled (the more usual case) and non boiled (raw).
c. semi-processed juices: products obtained by cold pressure or very rare by other treatments (diffusion, extraction, etc.) followed by the preservation.

Technical Processes for Preservation of Semi-processed Fruit Products

Preservation can be achieved by chemical means, by freezing or by pasteurization. The choice of preservation process for each individual case is a function of the semi-processed product type and the shelf life needed.

Chemical preservation. In many countries, in practice, this is carried out with sulphur dioxide, sodium benzoate, formic acid and, on a small scale, with sorbic acid and sorbates.

Preservation with sulphur dioxide is a widespread process because of its advantages: universal antiseptic action and very economic application. The drawbacks of SO_2 are: SO_2 turn firms the texture of some fruit species (pomaces), desulphiting is not always complete and recolouring of red fruits is not always complete after desulphitation. Practical preservation dosage levels with SO_2 for about 12 months is 0.18-0.20% SO_2 (with respect to the product to be preserved). This level could be reduced to 0.09% SO_2 for 3 months and to 0.12% SO_2 for 6 months preservation.

The preservation with sulphur dioxide is in use mainly for "pulps" and for "purées-marks".

Chemical preservation can be performed from a practical point of view by the utilisation of 6% SO2 water solutions or by direct introduction of sulphur dioxide gas in the product (for "purées-marks"). The preparation of 6% SO_2 solutions is done by bubbling the gas from cylinders in cold water; from a 50 kg SO_2 compressed gas cylinder results 830 l of 6% SO2 solution.

These SO_2 solutions have to be stored in cool places, in closed receptacles and with periodic concentration control/check by titration or by density measurements approximate results.

Preservation with sodium benzoate has the following advantages: it does not firm up the texture and does not modify fruit colour. The disadvantages are: it is not a universal antiseptic, its action needs an acid medium and the removal is partial. Sodium benzoate is in use for "pulps" and for "purées-marks" but less for fruit juices.

Practical dosage level for 12 months preservation is 0.18-0.20 % sodium benzoate, depending on the product to be preserved. Sodium benzoate is used as a solution in warm water; the dissolution water level has to be at maximum 10% reported to semi-processed product weight. Formic acid preservation is performed mainly for semi-processed fruit juices at a dosage level of 0.2 % pure formic acid (100%). Formic acid is an antiseptic effective against yeasts, does not influence colour of products and is easily removed by boiling.

Formic acid could be diluted with water in order to insure a homogeneous distribution in the product to be preserved; water has to be at maximum 5 % of the product weight. Because of a potential effect of pectic substance degradation, formic acid is less in use for "pulps" and "purées-marks" preservation.

Sorbic acid used as potassium sorbate (easily water soluble) can be used for preservation of fruit semi-processed products at a dosage level of 0. 1% maximum. Advantages of sorbates are: they are completely harmless and without any influence on the organoleptic properties of semi-processed fruit products.

Preservation by pasteurization. As fruit has a low pH, preservation of semiprocessed fruit products could also be performed by pasteurization (heat treatment step at maximum temperature of 100° C), the length of this step varying with the size of the receptacles.

The advantages of this type of treatment are: hygienic process, which assure a long term preservation; the disadvantages are: need for air tight receptacles, and pectic substances could begin to deteriorate if the thermal treatment is too long.

Thermal preservation of fruit semi-processed products could also be done by a "selfpasteurization": very hot semi-processed products are filled into receptacles (e.g. metal cans) which are sealed and then inverted in order to sterilise the air which goes through the hot fruit mass.

Preservation by freezing. This is done on an industrial scale in some countries and can be done with or without sugar addition. The advantages of this process are: absence of added substances; very good preservation of quality of fruit constituents (pectic substances, vitamins, etc.) and good preservation of organoleptic properties (flavour, taste, colour). Freezing is done at about -20 to -30° C and storage at -10 to -18° C.

Freezing is applied mainly to semi-processed fruit products aimed at very high quality and high cost finished products.

Technological flow-sheet for semi-processed fruit "pulps": chemical preservation.

Sorting is needed in order to remove sub-standard fruit (with moulds, with diseases, etc.) and all foreign bodies.

Washing is obligatory in order to remove all impurities which cannot be eliminated at the processing step in finished products.

Coring and Cutting, mainly for pomace fruits, has as main objective a better utilisation of preservation "space" in receptacles and is not mandatory; this will be defined by customer/supplier agreements/standards. This operation is preferably performed by mechanical means.

Preservation is carried out with the 6% SO2 solution which is added to the prepared fruits (placed in bulk in receptacles) in the quantity needed to obtain the preservation dosage level. For a better/ homogeneous preservative distribution, the initial 6% SO2 solution could be diluted with water; however, the diluted solution (which will be filled in receptacles) has to be at a dosage level of less than 10% of the semi-processed product weight. For some soft fruit, especially strawberries, preservation is done with a mix of 6% SO_2 solution and calcium bisulphite solution (containing also 6% SO_2).

Preparation of calcium bisulphite solution is done by the introduction of 30 kg of CaO in 1 m^3 SO_2 solution and mixing up to clarification. The resulting solution is mixed with the initial 6% SO2 solution, generally in a 1:1 ratio, but the ratio can be adapted to the fresh fruit texture. Firming of soft fruit texture by this treatment is based on the formation of calcium pectate with pectic substances from fruit tissues. In the case of sodium benzoate, formic acid or potassium sorbate, the dosage levels to be used are as indicated above with the rule that it is not allowed to add more that 10% liquid in receptacles on the prepared fruits.

Preservation by pasteurization or "self-pasteurization" will need as additional steps: a) boiling with a minimum water addition (maximum 10%); b) filling of receptacles; c) hermetic closing followed by d) pasteurization or "self-pasteurization".

Some general technical data for the preparation of chemical preserved semi-processed fruit "pulps" are seen in Table below.

Table: *General technical processing data for semi-processed fruit "pulps"*

Fruit species	Preliminary operations	Preservation means
Apples, pears, quinces	Sorting, washing, coring, cutting	Sulphur dioxide
Prunes, peaches, wax cherries, apricots	Sorting, washing, stone removal (pitting)	Sulphur dioxide or sodium benzoate
Cherries	Sorting, washing	Sulphur dioxide or sodium benzoate, sometimes with calcium bisulphite addition
Strawberries	Sorting, washing	Sulphur dioxide in mix with calcium bisulphite
Wild berries	Sorting, washing	Sulphur dioxide or sodium benzoate; in some cases with calcium bisulphite addition

Technological Flow-sheet for Semi-processed "Purée-marks"

The general technological flow-sheet includes the following operations:

Sorting and Washing are obligatory and are carried out in a similar manner as for "pulps". Heat Treatment/Boiling is needed in order to soften the fruit tissues before refining. For some fruits as strawberry and wild berries, this step is not done and fruits are refined "raw" in order to preserve their flavour. Pulping is performed with specific equipment - refiners or pulpers - which eliminate seeds, pits and other non edible parts (peels, cores, etc.).

Preservation is carried out by chemical means, by freezing or by pasteurization.

The general technical/processing data for the manufacture of "purée-marks" are seen in Table below.

Table: *General technical processing data for semi-processed "purées - marks"*

Fruit species	Preliminary operations	Preservation means
Apples, pears, quinces	Sorting, washing, boiling, refining	Sulphur dioxide; in less frequent cases formic acid or sodium benzoate
Prunes, peaches, wax cherries, apricots,	Sorting, washing, boiling, refining	Sulphur dioxide, formic acid or sodium benzoate; cherries freezing with or without sugar; self-pasteurization
Strawberry, wild berries	Sorting, washing, refining	Chemical preservation, freezing or self pasteurization

Fruit Sugar Preserves Technology; Jams, Jellies, Marmalade, Fruit Paste

As a overall rule of thumb, a sugar concentration of about 60% in finished or processed fruit products generally insures their preservation.

Preservation is not only determined by the osmotic pressure of sugar solutions but also by the water activity values in the liquid phase, which can be lowered by sugar addition; and by evaporation down to 0.848 aw; this value however does not protect products from mould and osmophile yeast attack. Maximum saccharose concentration that can be achieved in the liquid phase of the product is 67.89%; however higher total sugar quantities (up to 70-72%) found in products are explained by an increased reducing sugar solubility resulting from saccharose inversion.

Jams

The preservation of fruit by jam making is a familiar process carried out on a small scale by housewives in many parts of the world. Factory jam making has become a highly complex operation, where strict quality control procedures are employed to ensure a uniform product, but the manufacturing operations employed are in essence the same as those employed in the house. Fresh or pre-cooked fruit is boiled with a solution of cane or beet sugar until sufficient water has been evaporated to give a mixture which will set to a gel on cooling and which contains 32-34% water.

Gel formation is dependent on the presence in the fruit of the carbohydrate pectin, which at a pH of 3.2 - 3.4 and in the presence of a high concentration of sugar, has the property of forming a viscous semi-solid. During jam boiling, all micro-organisms are destroyed within the product, and if it is filled hot into clean receptacles which are subsequently sealed, and then inverted so that the hot jam contacts the lid surface, spoilage by micro-organisms will not take place during storage. The composition of jam made from stone fruit and berry fruit is shown in Table below. About 30% of the vitamin C present in fresh fruit is destroyed during the jam-making process, but that which remains in the finished product is stable during storage.

The high moisture content of jam (equivalent to an equilibrium relative humidity of about 82%) makes it susceptible to mould damage once the receptacle has been opened and exposed from some time to the air. No problems of microbiological spoilage are likely to arise in the canned product during storage.

Table: *Composition of some fruit jams*

Type of Jam	Moisture content %	Sugar (as invert sugar, %)	Vitamin C mg/100 g
Jam made from berry fruits: strawberry, raspberry, etc.	29.8	69.0	10 - 25
Jam made from stone fruits: apricot, peach, etc.	29.6	69.3	10 - 35

Source: FAD/WFP, 1970

Marmalade

This sugar preserve is defined as "semisolid or gel-like product prepared from fruit ingredients together with one or more sweetening ingredients and may contains suitable food acids and food pectins; the ingredients are concentrated by cooking to such a point that the TSS - Total Soluble Solids - of the finished marmalade is not below 65%".

Fruit Paste

Fruit paste is a product obtained in the same way as special non-gelified fruit marmalade but with a lower water content - about 25% TSS in fruit paste.

Lowering water content could be achieved by continuing boiling of the product or by drying the product by natural or artificial drying. An example of paste without sugar is the sun dried apricot or prune paste.

General procedure for the preparation of jams, jellies and marmalade:

a. Boil the pulp or the juice (with water when necessary)
b. Add the pectin
   * to the batch while stirring very vigorously
   * Pectin which has previously been mixed with 5 times its weight in sugar taken from the recipe)
c. Boil for about 2 minutes to assure a complete dissolution
d. Add the sugar while keeping the batch boiling
e. Boil down quickly to desired Brix
f. Add the acid (usually citric acid) and remove the froth
g. Fill hot into the (previously cleaned) jars and close
h. Invert the jars for three minutes to pasteurize the cover.

* Note: the pectin in solution can also be added at the end of the step (e) and has to be prepared as follows: use a strong blender. For one litre of water add slowly into the blender 25 g of pectin mixed with 100 g of sugar taken from the recipe.

Basic Recipes

The following recipes must be considered only as guidelines because the composition of the fruit can vary; also the taste of the consumers varies concerning the consistency, the sweetness and acidity.

Before starting to make jam it is important to know the yield to settle the question on containers. The calculation is made as follows:

$$\frac{\text{Total Soluble Solids content of all the raw ingredients}}{\text{Percentage of Total Soluble Solids in the final product}} \times 100$$

In these basic recipes it is assumed that the fruits are poor in pectin content.

Recipe 1. Fruit: sugar = 50:50; desired Brix in the finished product is 68.

Soluble Solids:

10 kg of fruit at 10% TSS	1.000 kg
10 kg of sugar	10.000 kg
60 g of pectin (grade 200)	0.060 kg
55 g of citric acid	0.055 kg
	11.115 kg

$$\text{Yield} = \frac{11.115 \times 100}{68} = 16.4 \text{ Kg}$$

Recipe 2. Fruit: sugar = 45:55; desired Brix in the finished product is 68.

Soluble Solids

d10 kg of fruit at 10% TSS	1.000 kg
2.5 litre of water	-
12.2 kg of sugar	12.200 kg
65 g of pectin grade 200	0.060 kg
60 g of citric acid	0.060 kg
	13.325 kg

$$\text{Yield} = \frac{13.325 \times 100}{68} = 19.6 \text{ Kg}$$

Recipe 3. Fruit: sugar = 40:60; desired Brix in the finished product is 68.

Soluble Solids:

d10 kg of fruit at 10% TSS	1.000 kg
3.3 litre of water	-
15 kg of sugar	15.000 kg

85 g of pectin grade 200	0.085 kg
80 g of citric acid	0.080 kg
	16.165 kg

$$\text{Yield} = \frac{16.165 \times 100}{68} = 23.8 \text{ Kg}$$

Various factors must be taken into account:

1. Size of the container: the quantity of pectin indicated in the recipes is valid for containers of 1 kg or less.

If container capacity is between:	*Increase pectin by:*
1 kg and 2.5 kg	5%
2.5 kg and 5.0 kg	10%
5.0 kg and 10.0 kg	20%
10.0 kg and 20.0 kg	30%

2. Finishing point: the quantity of pectin is given for a final Brix - Total Soluble Solids (TSS) of 68%.

If the final Brix is 66 increase the pectin by 5%

" " is 65 " " " by 10%

is 64 " " " by 15%

" " is 62 " " " by 20%

" " is 60 " " " by 30%

3. Acidic taste. If the product is too acid, replace the citric acid by tartaric acid (63% of the amount of citric acid).

4. Formation of clots: If batch clots, it is probably due to the pH being too low or the or TSS being too high; correct accordingly.

5. Formation of liquid at the surface: if liquid forms on the surface, it is probably due to too low a pH or too low pectin content.

6. Crystallisation: a) if liquid forms on the surfaces, then the pH is too low; reduce the acid content; b) if liquid does not form on the surface, then TSS or pH is too high.

7. Formation of mould: the TSS is probably below 68 deg. Brix. The filling may have been done at a low temperature. If the containers are large, wait until they are cold before closing.

8. Wrong batch: dilute the jam with water to 30% TSS; cook briefly. Add this diluted jam to a new batch but in a ratio not exceeding 10%.

Processing of Pineapple-papaya Jam

The fruit should be prepared as per previous instructions. For pineapples, the ends are removed and discarded; the cores and outer parts of the fruits are also removed. The fruit cylinders obtained are pulped through a special extractor (Fitzpatrick communiting machine) equipped with a 0.40-in screen sieve; the pulp thus obtained is used for making jam.

The papaya are prepared by hand-peeling the fruit; the fruit is then halved and the seeds removed. It is then pulped in the communiting machine using a 0.40-in screen sieve. When ginger root is used as flavouring, it is peeled and macerated in a Kenwood blender to a very fine consistency.

A typical formula for a pineapple-papaya jam (50:50 ratio) with ginger flavouring is given as follows:

Pineapple pulp	25.0
Papaya pulp	25.0 pounds
Cane sugar	50.0
Apple pectin (150 grade)	6.0 ounces
Citric acid	6.4
Fresh ground ginger	7.5

Processing is carried out in the following way:

The weighed fruit pulp is placed in a stainless steel steam-jacketed kettle and heated to about 110°F under constant stirring.

When the product reaches this temperature, the heat is turned off. The pectin (mixed in about ten times its weight with some of the weighed sugar), is then mixed into the fruit pulp, stirring constantly in order to prevent the pectin from clotting.

When the pectin has dissolved, the remainder of the sugar is added and dissolved completely in the mixture. The heat is then turned on and the jam mixture is stirred constantly until it starts boiling vigorously. During the remainder of the cooking, the product is stirred occasionally. Near the finishing point (approximately 221° F), the citric acid and the ginger (if it is used) are also added.

Determination of the finishing point is done by removing samples at intervals, cooling, and reading the soluble solids by means of a refractometer equipped with a Brix scale. After the jam reaches the proper Total Soluble Solids content, the heat is turned off and the surface scum/foam is removed.

The jam then is quickly put into receptacles which have been cleaned and sterilised with boiling in water for 30 minutes. The filling operation is done rapidly in order to prevent the temperature of the jam from falling below 190° F. After filling, sterilised lids (boiled for 30 minutes in water) are placed on the receptacles and they are then sealed.

After this operation the receptacles are inverted for about 3 minutes to insure that the lids are sterilised. The receptacles are then placed upright. At this stage it is not necessary to do any further processing, therefore the receptacles are cooled in running cold water until they reach a temperature slightly above room temperature. They are then dried in air and labelled.

Evaluation of Finished Products

During production at medium/large scale, it is recommended that quality controls be performed during manufacturing. After ten weeks of storage at room temperature it is recommended that an examination of finished products be performed. The receptacles are opened and contents carefully emptied on to enamel trays without disturbing the formation of the jam.

The empty cans (if metal cans were used) are then inspected for signs of corrosion. Factors other than flavour include colour, appearance, syrup separation, firmness and spreading quality. For flavour, jam is tested on pieces of bread. Samples are taken for measurement of pH (with a glass electrode pH meter) and Total Soluble Solids (with a refractometer equipped with a Brix scale). This evaluation enables to have a quality check during product shelf life and to obtain data needed for necessary improvements of future productions.

For pineapple-papaya jam, products made with 30% pineapple and 70% papaya with added ginger has the highest score for flavour.

The use of plain tin cans causes corrosion problems which is not the case when acid resistant lacquer cans are used.

Pineapple Jam Making

1. Boil 40 lb. of pulp and 12 lb. of water.
2. Add 225 g of pectin to the batch while stirring rapidly.
3. Boil for about 90 sec to assure complete dissolution.
4. Add 60 lb. of sugar gradually if possible in several portions, while keeping the batch boiling.

5. Boil down quickly to 69 deg. Brix (223 deg. F).
6. Take off steam, remove foam.
7. Add 300 cc citric acid solution 50%.
8. Fill hot (180 deg. F).
9. Invert receptacles for 3 minutes.

Check each batch for Brix 68-70 deg.; acidity/pH = 3.2 +/- 0.2.

Evaluate: absence of defects; colour; flavour; consistency.

Various fruit: sugar ratios can be manufactured; some basic recipes are as follows:

	Ratio 50:50	*Ratio 45:55*	*Ratio 40:60*
Fruit	55 lb	49.5 lb	44 lb
Water	-	11 lb	13.2 lb
Sugar	55 lb	60.5 lb	66 lb
Pectin (150 grade)	225 g	237.5 g	250 g
Citric acid (50% sol.)	300 cc	320 cc	335 cc

Fruit Juice Technologies

Fruit juices are products for direct consumption and are obtained by the extraction of cellular juice from fruit, this operation can be done by pressing or by diffusion.

For the purpose of this document, the technology of fruit juice processing will cover two finished product categories:

- juices without pulp ("clarified" or "not clarified");
- juices with pulp ("nectars").

We will also define as

a. "natural juices" products obtained from one fruit; and
b. "mixed juices" products obtained from the mix of two or three juices from different fruit species or by adding sugar.

Juices obtained by removal of a major part of their water content by vacuum evaporation or fractional freezing will be defined as "concentrated juices".

Technological Steps for Processing of Fruit Juices Without Pulp

Fruit juices must be prepared from sound, mature fruit only.

Soft fruit varieties such as grapes, tomatoes and peaches should only be transported in clean boxes which are free from mould and bits of rotten fruit.

Washing: fruit must be thoroughly washed. Generally, fruit will be submitted to a pre-washing before sorting and a washing step just after sorting.

Sorting: removal of partially or completely decayed fruit is the most important operation in the preparation of fruit for production of first quality fruit juices; sorting is carried out on moving inspection belts or sorting tables.

Crushing/Grinding/Disintegration Step is applied in different ways and depends on fruit types:

Crushing for grapes and berries;

Grinding for apples, pears;

Disintegration for tomatoes, peaches, mangoes, apricots.

This processing step will need specific equipment which differs from one type of operation to another.

Enzyme Treatment of crushed fruit mass is applied to some fruits by adding 2-8% pectolitic enzymes at about 50° C for 30 minutes.

This optional step has the following advantages: extraction yield will be improved, the juice colour is better fixed and finished product taste is improved. However, for fruit which is naturally rich in pectic substances, this treatment makes the resulting "exhausted" material useless for industrial pectin production.

Heating of crushed fruit mass before juice extraction is an optional step used for some fruit in order to facilitate pressing and colour fixing; at same time, protein coagulation takes place.

Pressing to extract juice. Diffusion is an alternative step for juice extraction and can be carried out discontinuously or in batteries at water temperature of about 80-85 ° C.

Juice Clarifying can be performed by centrifugation or by enzyme treatment. Centrifugation achieves a separation of particles in suspension in the juice and can be considered as a pre-clarifying step. This operation is carried out in centrifugal separators with a speed of 6000 to 6500 RPM.

Enzyme clarifying is based on pectic substance hydrolysis; this will decrease the juices' viscosity and facilitate their filtration. The treatment is the addition of pectolitic enzyme preparations in a quantity of 0.5 to 2 g/l and will last 2 to 6 hours at room temperature, or less than 2 hours at 50° C, a temperature that must not be exceeded.

The control of this operation is done by checking the decrease in juice viscosity. Sometimes, the enzyme clarifying is completed with the step called "sticking" by the addition of 5-8 g/hl of food grade gelatine which generates a flocculation of particles in suspension by the action of tannins.

Filtration of clarified juice can be carried out with kieselgur and bentonite as filtration additive in press-filters (equipment).

De-tartarisation is applied only to raisin juice and is aimed to eliminate potassium bitartrate from solution. This step can be performed by the addition of 1% calcium lactate or 0.4% calcium carbonate.

Pasteurization of juice can be done for temporary preservation (pre-pasteurization) and in this case this operation is carried out with continuous equipment (heat exchangers, etc.); warm juice is stored in drums or large size receptacles (20-30 kg). Pasteurization conditions are at 75°C in continuous stream.

Pasteurization of bottled juice is then carried out just before delivery to the market; this is performed in water baths at 75° C until the point where the juice reaches 68° C. In cases when the final pasteurization is done without pre-pasteurization and temporary storage, modern methods use a rapid pasteurization followed by aseptic filling in receptacles.

Rapid pasteurization conditions are as follows: temperature about 80° C, over 10-60 sec., followed by cooling; all operations are carried out in continuous stream.

Preservation under CO_2 pressure may be done at a concentration of 1.5% CO_2 under a pressure of 7 kg/cm². At the distribution step, proceed at CO_2 decompression and the juice is then submitted to a sterilising filtration and aseptic filling in receptacles.

Preservation by freezing is carried out at about -30° C, after a preliminary de-aeration; storage is at -15 to -20° C.

Production of concentrated juices by evaporation is performed under vacuum (less than 100 mm Hg residual pressure) up to a concentration of 65-70% total sugar which assures preservation without further pasteurization. Modern evaporation installations recover flavours from juices which are then reincorporated in concentrated juices.

Additional operations for juice manufacturing are the vacuum de-aeration and mixing with other fruit juices or with sugar.

For the production of non clarified juices the centrifugation is the only specific step, enzyme clarifying and subsequent filtration being eliminated.

The optimum sugar/acid ratio for the majority of fruit, mainly for pomaces, is 10/1 to 15/1.

Fruit which is rich in carotenoids (apricots, peaches, etc.) is only processed as juices with pulp ("nectars").

Technological Flow-sheet for Fruit Juices with Pulp ("Nectars")

This process is divided at industrial scale in two categories of operations:

a. processing for obtaining juices;

b. juice conditioning for preservation.

Operations in the first category are differ according to the type of fruit which to be processed. Pomaces (apples, pears) are washed and sorted and then crushed in a colloid mill; fruit purée is then passed through a screw type heating equipment where direct steam is used as a source of heat. Warm fruit mass is treated in a pulper with a 2 mm screen and then through an extractor similar with the equipment used for tomato juice.

Stone fruits (apricots, peaches, cherries, etc.) after washing and sorting are submitted to steam in a continuous heater, then the warm fruit mass is passed through a pulper and then an extractor (as mentioned above). Berries (strawberry, wild berries, etc.) are washed, sorted and then crushed, preheated and then introduced in extractor.

In order to avoid browning and undesirable taste modifications it is usual to add about 0.05% ascorbic acid.

Second category type of operations are similar for all fruit species. Partial elimination of cellulose is achieved with a continuous centrifugal separator; the resulting juice is then processed in order to adjust sugar and acid content for viscosity.

Sugar (about 8-10%) is added as a syrup (in water or in the juice of same fruit obtained by pressure). Acidity is adjusted with citric or tartaric acid. The adjusted juice is then deaerated under vacuum at about 40° C. This step aims at avoiding oxidative reactions and vitamin C loss reduction.

An important subsequent step is an intensive homogenisation (under pressure at 150-180 A) in order to obtain particles with dimensions below 100. The homogenised juice obtained is then continuously pasteurized in plate heat exchanger equipment at a temperature of about 130° C, cooled down to about 90° C and aseptically packed in receptacles.

The principal characteristics of fruit "nectars" are uniformity and stability of the content provided by the advanced disintegration of fruits. Stability can be obtained by increasing product viscosity by adding pectin for fruit which is deficient in this component. In order to avoid "separation", intensive homogenisation is carried out as described above.

Fruit "nectars" contain all the important components of the original fruit and to a large extent maintain their taste and flavour. The sugar/ acidity (as citric acid) ratio is to a large extent determined by the type of fruit and the correction applied; for example, this ratio is 30 for apricots, 40 for peaches, 160 for pears, etc.

Banana and Plantain Processing Technologies

Products: Uses and Dietary Significance

Most of the world's bananas are eaten either raw, in the ripe state, or as a cooked vegetable, and only a very small proportion are processed in order to obtain a storable product. This is true both at a traditional village level with both dessert and cooking bananas and when considering the international trade in dessert bananas.

In general, preserved products do not contribute significantly to the diet; however, in some localised areas the products are important in periods when food are scarce.

Probably the most widespread and important product is flour preparation from unripe banana and plantains by sun-drying. In Uganda, dried slices known as "mutere" are prepared for storage from green bananas, the dried slices being either used directly for cooking or after grinding into a flour. "Mutere" is used chiefly as a famine reserve and does not feature largely in the diet under normal conditions.

In Gabon, plantains are sometimes made into dried slices which can be stored and used on long journeys, and plantains are used in Cameroon to prepare dried pieces which are stored and ground as needed into flour for use in cooking a paste known as "fufu". Dried green banana slices are also used in parts of South and Central America and West Indies for preparing flour. The other nutritionally important product is beer which is a major product in Uganda, Rwanda and Burundi where green banana utilisation is particularly high.

Preservation Methods and Processes

Drying. - Both ripe and unripe bananas and plantains are normally peeled and sliced before drying, although banana figs are sometimes prepared from whole ripe fruit. Sun drying is the most widespread technique where the climate is suitable but drying in ovens or over fires is also practiced. In west Africa, plantains are often soaked and sometimes parboiled before drying.

The slices of unripe fruit are normally spread out on bamboo frameworks; or a cemented area; or on a mat; or on a swept-bare patch of earth; or on a roof; or sometimes on stones outcrops or sheets of corrugated iron. Oven-drying of ripe bananas is practiced in Polynesia as a mean of preserving the fruits, which are then wrapped in leaves and bound tightly to store until needed. In East Africa a method has been reported that involves drying the peeled bananas on a frame placed over a fire for 24 hr before drying in the sun, to accelerate the process.

Potential for Scaling up of Traditional Processes to Industrial Level

Many banana products are now produced on an industrial scale, including the traditional banana figs and flour, and the processing techniques are described below. One of the main problems encountered has been the susceptibility of banana products to flavour deterioration and discoloration and in the past many products reaching the market have been of poor quality.

A great deal of research has been directed to overcoming these problems, although however good the resultant products are they cannot compare in flavour and other characteristics with the fresh banana fruit. Indeed, an important constraint on the large-scale development of banana processing is the lack of demand for banana products since the fresh fruit is available throughout the year in most parts of the tropical world.

The production of beer from banana and plantains has not been scaled up to an industrial level, and while an important product in localised areas of tropical Africa, the market is rapidly declining in favour of European-type brews produced locally.

Industrial Processing

The main commercial products made from bananas are canned or frozen purée, dried figs, banana powder, flour, flakes, chips (crisps), canned slices and jams.

Banana products can be divided roughly into two types - those for direct consumption, such as figs, and those for use in food manufacturing industry, for example purées and powder.

Banana figs, or fingers as they are sometimes known, are usually whole, peeled fruit carefully dried so as to retain their shape, although sometimes the fruit is sliced or halved to facilitate drying. Banana and plantain chips (crisps) are thinly sliced pieces of fruit fried in oil and eaten as a snack like potato chips (crisps).

The main use of canned slices is in tropical fruit salads. Banana flakes are used as a flavouring or in breakfast cereals. Banana purée find use mainly in the production of baby foods. Banana flour is said to be highly digestible and is used in baby and invalid foods, but can also be used in the preparation of bread and beverages.

Banana powder is used chiefly in the baking industry for the preparation and fillings for cakes and biscuits and is also used for invalid and baby foods.

Processing Technology

In general, to obtain a good-quality product from ripe-bananas the fruit is harvested green and ripened artificially under controlled conditions at the processing factory. After ripening, the banana hands are washed to remove dirt and any spray residues, and peeled. Peeling is almost always done by hand using stainless steel knives, although a mechanical peeler for ripe bananas has been developed, capable of peeling 450 Kg of fruit per hour (Banana Bulletin, 1974).

The peeling of unripe bananas and plantains is facilitated by immersing the fruit in hot water. For example, immersion in water at 70-75 ° C for 5 min. has been suggested as an aid for peeling green bananas for flour production, while the peeling of green bananas for freezing has been facilitated by immersion in water at 93° C for 30 min.

Banana Figs

Fully ripe fruits with a sugar content of about 19.5% are used and are treated with sulphurous acid after peeling, then dried as soon as possible afterwards. Various drying systems have been described using temperatures between 50 and 82° C for 10 to 24 hr to give a moisture content ranging from 8 to 18% and a yield of dried figs of 12 to 17% of the fresh banana on the stem.

One factory in Australia uses a solar heat collector on the roof to augment the heat used for drying bananas. Bananas can also be dried by osmotic dehydration, using a technique which involves drawing water from 1/4-in. thick banana by placing them in a sugar solution of 67 to 70 deg. Brix for 8 to 10 hr. followed by vacuum-drying at 65 to 70° C, at a vacuum of 10 mm Hg for 5 hr. The moisture content of the final products is 2.5% or less, much lower than that achieved by other methods.

Banana Purée

Banana purée is obtained by pulping peeled, ripe bananas and then preserving the pulp by one of three methods: canning aseptically, acidification followed by normal canning, or quick-freezing. The bulk of the world's purée is processed by the aseptic canning technique. Peeled, ripe fruits are conveyed to a pump which forces them through a plate with 1/4-in. holes, then onto a homogeniser, followed by a centrifugal de-aerator, and into a receiving tank with 29in. vacuum, where the removal of air helps prevent discoloration by oxidation.

The purée is then passed through a series of scraped surface heat exchangers where it is sterilised by steam, partially cooled, and finally brought to filling temperature. The sterilised purée is then packed aseptically into steam-sterilised cans which are closed in a steam atmosphere.

Banana Slices

Several methods for canning of banana slices in syrup are used. Best-quality slices are obtained from fruit at an early stage of ripeness. The slices are processed in a syrup of 25 deg. Brix with pH about 4.2, and in some processes calcium chloride (0.2%) or calcium lactate (0.5%) are added as firming agents. A method for producing an intermediate-moisture banana product for sale in flexible laminate pouches has been developed. Banana slices are blanched and equilibrated in a solution containing glycerol (42.5%), sucrose (14.85%), potassium sorbate (0.45%), and potassium metabisulphite (0.2%) at 90 deg. C for 3 min. to give a moisture content of 30.2%.

Banana Powder

In the manufacture of banana powder, fully ripe banana pulp is converted into a paste by passing through a chopper followed by a colloid mill. A 1 or 2 % sodium metabisulphite solution is added to improve the colour of the final product. Spray- or drum-drying may be used, the latter being favoured as all the solids are recovered.

A typical spray dryer can produce 70 kg powder per hour to give yields of 8 to 11% of the fresh fruit, while drum-drying gives a final yield of about 13% of the fresh fruit. In the latter method the moisture content is reduced to 8 to 12 % and then further decreased to 2 % by drying in a tunnel or cabinet dryer at 60° C.

Banana Flour

Production of flour has been carried out by peeling and slicing green fruit, exposure to sulphur dioxide gas, then drying in a counter-current tunnel dryer for 7 to 8 hr. with an inlet temperature of 75° C and outlet temperature of 45° C, to a moisture content of 8%, and finally milling.

Banana Chips (Crisps)

Typically, unripe peeled bananas are thinly sliced, immersed in a sodium or potassium metabisulphite solution, fried in hydrogenated oil at 180 to 200° C, and dusted with salt and an antioxidant.

Alternatively, slices may be dried before frying and the antioxidant and salt added with the oil. Similar processes for producing plantain chips have been developed.

Banana Beverages

In a typical process, peeled ripe fruit is cut into pieces, blanched for 2 min. in steam, pulped and pectolytic enzyme added at a concentration of 2 g enzyme per 1 kg pulp, then held at 60 to 65° C and 2.7 to 5.5 pH for 30 min.

In a simpler method, lime is used to eliminate the pectin. Calcium oxide (0.5%) is added to the pulp and after standing for 15 min. this is neutralised giving a yield of up to 88% of a clear, attractive juice. In another process banana pulp is acidified, and steam-blanched in a 28-in Hg vacuum which ensures disintegration and enzyme inactivation. The pulp is then conveyed to a screw press, the resulting purée diluted in the ratio 1:3 with water, and the pH adjusted by further addition of citric acid to 4.2 to 4.3, which yields an attractive drink when this is centrifuged and sweetened.

Jam

A small amount of jam is made commercially by boiling equal quantities of fruit and sugar together with water and lemon juice, lime juice or citric acid, until setting point is reached.

Product Stability and Spoilage Problems

All dried banana products are very hydroscopic and susceptible to flavour deterioration and discoloration, but this can be overcome to some extent by storing in moisture-proof containers and sulphiting the fruit before drying to inactivate the oxidases.

The dried products are also liable to attack by insects and moulds if not stored in dry conditions, although disinfestation after drying by heating for 1 hr to 80° C or by fumigation with methyl bromide ensures protection against attack. Banana powder is said to be stored for up to a year commercially and flakes have been stored in vacuum-sealed cans with no deterioration in moisture, colour or flavour for 12 months.

Banana chips tend to have a poor storage life and to become soft and rancid. However, chips treated with an antioxidant have been stored satisfactorily at room temperature in hermetically sealed containers up to 6 months with no development of rancidity.

Quality Control Methods

In general a good quality product is obtained if fruit is harvested at the correct stage of maturity and, where appropriate, ripened under controlled conditions. For example, in the case of banana figs, the fruit should be fully mature (sugar content of 19.5% or above) or the final product is liable to be tough and lacking in flavour. However, if over-ripe fruit is used, the figs tend to be sticky and dark in colour, so the fruit must be fully yellow but still firm.

For banana flour, which is prepared from unripe bananas, the fruit is harvested at three-quarters the full-ripe stage and is processed within 24 hr. prior to the onset of ripening. If less mature fruit is used, the flour tastes slightly astringent and bitter due to the tannin content. Bananas harvested between 85 and 95 days after the emergence of the inflorescence, with a pulp-to-peel ratio of about 1.7, were considered to be most suitable for the deep-fat frying.

Other criteria suggested for assessing maturity were beta-carotene and reducing sugar content, both of which increase with increasing maturity and pH which decreases as the fruit ripens, and these should be, respectively, about 2000 μg/100 g, less than 1.5% and 5.8 or above. Browning was found to occur if the sugar content was higher than 1.5%. The determination of crude fat in processed chips is also considered to be a necessary quality control measure.

It is important to remove all impurities prior to processing of products, and this is done by washing to remove dirt and spray residues and control on the processing line so that substandard fruit can be removed.

Preparation Methods for Fresh Bananas and Plantains

The main ways of preparing fresh bananas for consumption are boiling or steaming, roasting or baking and frying. Boiling followed by pounding into "fufu" is also widely adopted in certain areas of the tropics.

Boiling or Steaming

Plantains and bananas are often prepared simply by boiling in water, either in their peel or after peeling, and either ripe or unripe; if unripe, the fruit is scraped thoroughly after peeling to remove all traces of fibrous material. The boiled fruit is eaten alone or more usually accompanied by a sauce. This preparation technique is widely used in West Africa.

Roasting or baking. Unpeeled or peeled fruit, either ripe or unripe, is roasted simply by placing in the ashes of a fire or in an oven. This method is widely used in West Africa, East Africa and the South Pacific islands. For example, ripe plantains are placed unpeeled in an oven and when partly brown and tender, removed and peeled, then replaced in the oven and roasted evenly.

Frying. Ripe or unripe plantains or bananas are often peeled, sliced and cooked in oil, particularly in West Africa and in parts of South America and the West Indies. Similar products are also made in East Africa. Typically, ripe plantains are peeled, cut into slices or split lengthways, and fried in palm oil or with groundnut oil, the pieces being served either hot with a sauce or with fried eggs, or cold as a snack.

Pounding. Pounding is a process, used particularly in West Africa, for preparing most perishable staple food crops including plantains, cassava, yams and cocoyams to obtain a paste or dough known as "fufu" (also spelled "foofoo", "foutou", "foufou"). The plantains are peeled or boiled and peeled after boiling and pounded in a wooden mortar, the resulting paste normally being eaten with soup or a spiced sauce of meat and vegetables, but sometimes after wrapping in leaves and steaming.

Mango and Guava Processing Technologies

Mangoes are processed at two stages of maturity. Green fruit is used to make chutney, pickles, curries and dehydrated products. The green fruit should be freshly picked from the tree. Fruit that is bruised, damaged, or that has prematurely fallen to the ground should not be used. Ripe mangoes are processed as canned and frozen slices, purée, juices, nectar and various dried products. Mangoes are processed into many other products for home use and by cottage industry.

The mango processing presents many problems as far as industrialization and market expansion is concerned. The trees are alternate bearing and the fruit has a short storage life; these factors make it difficult to process the crop in a continuous and regular way. The large number of varieties with their various attributes and deficiencies affects the quality and uniformity of processed products.

The lack of simple, reliable methods for determining the stage of maturity of varieties for processing also affects the quality of the finished products. Many of the processed products require peeled or peeled and sliced fruit. The lack of mechanised equipment for the peeling of ripe mangoes is a serious bottleneck for increasing the production of these products.

A significant problem in developing mechanised equipment is the large number of varieties available and their different sizes and shapes. The cost of processed mango products is also too expensive for the general population in the areas where most mangoes are grown. There is, however, a considerable export potential to developed countries but in these countries the processed mango products must compete with established processed fruits of high quality and relatively low cost.

Green Mango Processing

Pickles. The optimum stage of maturity should be determined for each variety used to make pickles.

There are two classifications of pickles - salt pickles and oil pickles. They are processed from whole and sliced fruit with and without stones. Salt is used in most pickles.

The many kinds of pickles vary mainly in the proportions and kinds of spices used in their preparation. One basic recipe for the study of the preparation and storage of pickles in oil is as follows:

Mango pieces	250 g	Tumeric powder	2 to 4 g
Salt	60 g	Fenugreek seeds	2 to 4 g
Mustard powder	30 g	Bengal gram seeds	2 to 4 g
Chili powder	20 g	Gingelly oil	20 to 30 g

The ingredients are mixed together and filled into wide-mouthed bottles of 0.5 kg capacity. Three days later the contents are thoroughly mixed and refilled into the bottles. Extra oil is added to form a 1-2 cm layer over the pickles.

Chutney. The product is prepared from peeled, sliced or grated unripe or semi-ripe fruit by cooking the shredded fruit with salt over medium heat for 5 to 7 minutes, mixed and then sugar, spices and vinegar are added. Cook over moderate heat until the product resembles a thick purée, add remaining ingredients and simmer another 5 min. Cool and preserve in sterilised jars.

Spices usually include cumin seeds, ground cloves, cinnamon, chili powder, ginger and nutmeg. Other ingredients such as dried fruits, onions, garlic and nuts may be added.

Drying/dehydration. Immature fruit is peeled and sliced for sun-drying. The dried mango slices can be powdered to make a product called amchoo. The use of blanching, sulphuring and mechanical dehydration gives a product with better colour, nutrition, storability and fewer microbiological problems.

Ripe Mango Processing

Purée. Mangoes are processed into purée for re-manufacturing into products such as nectar, juice, squash, jam, jelly and dehydrated products. The purée can be preserved by chemical means, or frozen, or canned and stored in barrels. This allows a supply of raw materials during the remainder of the year when fresh mangoes are not available.

It also provides a more economical means of storage compared with the cost of storing the finished products, except for those which are dehydrated, and provides for more orderly processing during peak availability of fresh mangoes.

Mangoes can be processed into purée from whole or peeled fruit. Because of the time and cost of peeling, this step is best avoided but with some varieties it may be necessary to avoid off-flavours which may be present in the skin. The most common way of removing the skin is hand-peeling with knives but this is time-consuming and expensive. Steam and lyepeeling have been accomplished for some varieties.

Several methods have been devised to remove the pulp from the fresh ripe mangoes without hand-peeling. A simplified method is as follows: the whole mangoes were exposed to atmospheric steam for 2 to 2 1/2 min in a loosely covered chamber, then transferred to a stainless steel tank.

The steam-softened skins allowed the fruit to be pulped by a power stirrer fitted with a saw-toothed propeller blade mounted 12.7 to 15.2 cm below a regular propeller blade. The pulp is removed from the seeds by a continuous centrifuge designed for use in passion fruit extraction. The pulp material is then passed through a paddle pulper fitted with a 0.084 cm screen to remove fibre and small pieces of pulp.

Mango purée can be frozen, canned or stored in barrels for later processing. In all these cases, heating is necessary to preserve the quality of the mango purée. In one process, purée is pumped through a plate heat exchanger and heated to 90°C for 1 min and cooled to 35° C before being filled into 30 lb tins with polyethylene liners and frozen at -23.50 C.

In an other process, pulp is acidified to pH 3.5, pasteurized at 90°C, and hot-filled into 6 kg high-density bulk polyethylene containers that have been previously sterilised with boiling water. The containers are then sealed and cooled in water. This makes it possible to avoid the high cost of cans.

Wooden barrels may be used to store mango pulp in the manufacture of jams and squashes. The pulp is acidified with 0.5 to 1.0% citric acid, heated to boiling, cooled, and SO2 is added at a level of 1000 to 1500 ppm in the pulp. The pulp is then filled into barrels for future use.

Slices. Mango slices can be preserved by canning or freezing, and recent studies have shown the feasibility of pasteurized-refrigerated and dehydro-canned slices. The quality of the processed product in all of these procedures will be dependent upon selection of a suitable variety along with good processing procedures. Thermal process canning of mango slices in syrup is the most widely used preservation method.

Beverages : The commercial beverages are juice, nectar and squash. Mango nectar and juice contain mango purée, sugar, water and citric acid in various proportions depending on local taste, government standards of identity, pH control, and fruit composition of the variety used. Mango squash in addition to the above may contain SO_2 or sodium benzoate as a preservative. Other food grade additives such as ascorbic acid, food colouring, or thickeners may be used in mango beverages. A short description of finished products found in literature is as follows:

- mango juice: prepared by mixing equal quantities of pulp (purée) and water together and adjusting the total soluble solids (TSS) and acidity to taste (12 to 15% TSS and 0.4 to 0.5% acidity as citric acid);
- mango nectar containing 25% purée can be prepared using the following procedure.

		Brix of purée	
Nectar components	15°	17°	20°
Purée	100	100	100
Sugar	45	43	40
Water	255	257	260

Commercial processing conditions may require the use of a preservative. The pH is adjusted to approximately 3.5 by adding citric acid as a 50% solution.

The time of heat processing will vary with filling temperature, can size and viscosity of the juice or nectar.

Mango squash may be prepared according to flow-sheet described below; the finished product may contain 25% juice, 45% TSS and 1.2 to 1.5% acidity and may be preserved with sulphur dioxide (350ppm) or sodium benzoate (1000 ppm) in glass bottles.

Mango squash simplified flow-sheet.

Ingredients		
Mango pulp	900	900
Sugar	900	1100
Citric acid	18	15
Water	900	900

Mangoes are washed, stored, peeled with stainless steel knives. The pulp is prepared by using a pulper with fine sieve (0.025-in); Sugar is mixed with water and citric acid = syrup; The pulp is added to the syrup and mixed well; The mixture is strained trough cloth; The squash is heated at 85° C and bottles are filled and closed.

For additional heat treatment bottles may need to be maintained at a product temperature of 80°C for 30 minutes if the product is to be processed without preservatives. The bottles are then left to cool in water and stored at room temperature.

Two negative points must be avoided: presence of air bubbles (which is a source of quick deterioration) and separation of squash solids (giving an undesirable appearance). The means to avoid these two phenomena are described in the fruit juices section. A type of "squash type" beverage may also be manufactured with '/a pulp + '/a water + i/a sugar and pH adjusted to 3.7 by addition of citric acid. Using different sieve sizes affects the quality and reduces air bubbles to a certain extent but homogenisation and de-aeration of purée or squash seem to be important in order to avoid separation and air bubbles. The squash quality is evaluated on the basis of the following characteristics: pH, titrable acidity, soluble solids, ascorbic acid (by 2,6 dichlorophenol indophenol method), specific gravity.

Dried/dehydrated. Ripe mangoes are dried in the form of pieces, powders, and flakes. Drying procedures such as sun-drying, tunnel dehydration, vacuum-drying, osmotic dehydration may be used. Packaged and stored properly, dried mango products are stable and nutritious. One described process involves as pre-treatment dipping mango slices for 18 hr (ratio 1:1) in a solution containing 40° Brix sugar, 3000 ppm SO2, 0.2% ascorbic acid and 1% citric acid; this method is described as producing the best dehydrated product. Drying is described using an electric cabinet through flow dryer operated at 60° C. The product showed no browning after 1 year of storage.

Drum-drying of mango purée is described as an efficient, economical process for producing dried mango powder and flakes. Its major drawback is that the severity of heat preprocessing can produce undesirable cooked flavours and aromas in the dried product. The drum-dried products are also extremely hydroscopic and the use of in-package desiccant is recommended during storage.

Canning. This preservation technology is described in various technological flow-sheets in this bulletin.

Mango bar or "fruit leather" is presented in various flow-sheets.

Guava Processing Technologies

Guava purée. Guava purée is used in the manufacture of guava nectar, various juice drink blends and in the preparation of guava jam. The washed sound fruit is first passed through a chopper or slicer to break up the fruit and this material is fed into a pulper. The pulper will remove the seeds and fibrous pieces of tissue and force the reminder of the product through a perforated stainless steel screen. The holes in the screen should be between 0.033 and 0.045 in. The machine should be fed at a constant rate to ensure efficient operation.

The puréed material coming from the pulper is next passed through a finisher. The finisher is equipped with a screen containing holes of approximately 0.020 in. The finisher will remove the stone cells from the fruit and provide the optimum consistency to the product.

Perhaps the best way to preserve the guava purée is by freezing and the material passing through the finisher can be packaged and frozen with no further treatment. It is not necessary to heat the product to inactivate enzymes or for other purposes. The material can be frozen in a number of types of cartons and cans; however, a fibre box with a plastic bag inside is commonly used and is probably the less expensive.

It is also possible to can and heat process the guava purée and this can be accomplished by heating the purée to 195° F in an open double bottom kettle, filling into cans, closing the cans, inverting the cans for a few seconds, followed by cooling. Cans should be cooled rapidly to approximately 100° F before they are cased and stacked into warehouses.

Guava Juice and Concentrate

Guava juice can be used in the manufacture of a clear guava jelly or in various drinks. A clear juice may be prepared from guava purée that is depectinised enzymatically. About 0.1% pectin-degrading enzyme is mixed into the purée at room temperature; heating of the product at approximately 120° F will greatly speed the action of the enzyme. After 1 hr. clear juice is separated from the red pulp by centrifuging or by pressing in a hydraulic juice press. A batch-type or continuous-flow centrifuge can be used on the depectinised purée with no further treatment.

The clear juice after centrifuge or after press (and subsequent filtration) can be preserved by freezing or by pasteurization in hermetically sealed cans.

For shipment to overseas markets it may be advantageous to concentrate either the purée or the juice.

Recent Trends in Fruit and Vegetable Processing

New products. The number and variety of fruit and vegetable products available to the consumer has increased substantially in recent years. The fruit and vegetable industry has undoubtedly benefited from the increased recognition and emphasis on the importance of these products in a healthy diet.

Traditional processing and preservation technologies such as heating, freezing and drying together with the more recent commercial introduction of chilling continue to provide the consumer with increased choice. This has been achieved by new heating (e.g. UHT, microwave, ohmic) and freezing (e.g. cryogenic) techniques combined with new packaging materials and technologies (e.g. aseptic, modified atmosphere packaging).

The overall trend in new fruit and vegetable products is "added value", thus providing increased convenience to the consumer by having much greater variety of ready prepared fruit and vegetable products. These may comprise complete meals or individual components. The suitability of products and packages for microwave re-heating has been an important factor with respect to added convenience.

Table: *Numbers of new fruit and vegetable products*

	1990	***1991***	***1992 Jan–June***
Frozen vegetable products	66	95	21
Chilled vegetable products	76	81	78
Heat processed vegetables	51	60	38
Heat processed fruits	13	14	5
Fruit juices and drinks	73	83	46
Potato crisps	32	33	16

Source: C. Dennis (1993). New product development in the fruit and vegetable sector is most important in meeting the continued challenge of providing the consumer with choice and high quality products.

A Fresh Look at Dried Fruit

New fruit varieties and advance in drying technologies are putting a fresh twist on dried fruit applications. Fruits that have been introduced to the drying process include cranberries, blueberries, cherries, apples, raspberries and strawberries - not to mention the traditional mainstays of raisins, dates, apricots, peaches, prunes and figs.

Perceived as a "value-added" ingredient, dried fruit adds flavour, colour, texture and diversity with little alteration to an existing formula. The growing interest in ethnic cuisines in U.S.A. and the change to a more healthy way of eating, has also moved dried fruit considerably closer to the mainstream.

Found primarily in the baking industry, dried fruit is coming into its own in various food products, including entrees, side dishes and condiments. Compotes, chutneys, rice and grain dishes, stuffings, sauces, breads, muffins, cookies, deserts, cereals and snacks are all food categories encompassing dried fruit.

Since some dried fruit is sugar infused (osmotic drying), food processors can decrease the amount of sugar in formula - this is especially the case in baked products. Processors are making adjustments in moisture content of the dried fruit so that a varied range is available for different applications. An added bonus is dried fruits' shelf stability (a shelf life of at least 12 months). Dried fruit is more widely available in different forms, including whole dried, cut, diced and powders.

Citric acid and its use in fruit and vegetable processing. Citric acid may be considered as "Nature's acidulant". It is found in the tissues of almost all plants and animals, as well as many yeasts and moulds.

Commercially citric acid is manufactured under controlled fermentation conditions that produce citric acid as a metabolic intermediate from naturally-occurring yeasts, moulds and nutrients. The recovery process of citric acid is through crystallization from aqueous solutions.

Citric acid is widely used in carbonated and still beverages, to impart a fresh-fruit "tanginess". Citric acid provides uniform acidity, and its light fruity character blends well and enhances fruit juices, resulting in improved palatability. The amount of citric acid used depends on the particular desired flavour (e.g., High-acid: lemonade;

Medium-acid: orange, punch, cherry; Low-acid: strawberry, black cherry, grape).

Sodium citrate is often added to beverages to mellow the tart taste of high acid concentrations. It provides a cool, distinctive smooth taste and masks any bitter aftertaste of artificial sweeteners. In addition, it serves as a buffer to stabilise the pH at the desired level. The high water solubility of citric acid (181 g/100 ml) makes it an ideal additive for fountain fruit syrups and beverages concentrates as a flavour enhancer and microbial growth inhibitor (preferably at pH < 4.6).

In processed fruits and vegetables, citric acid performs the following functions:

a. It reduces heat-processing requirements by lowering pH: inhibition of microbial growth is a function of pH and heat treatment. Higher heat exposure and lower pH result in greater inhibition. Thus the use of citric acid to bring pH below 4.6 can reduce the heating requirements. In canned vegetables, citric acid usage is greatest in tomatoes, onions and pimentos. For tomato packs, the National Canners Association recommends a pH of 4.1 to 4.3. In general, 0. 1% citric acid will reduce the pH of canned tomatoes by 0.2 pH units.
b. Optimise flavour: citric acid is added to canned fruits to provide for adequate tartness. Recommended usage level is generally less then 0.15%.
c. Supplement antioxidant potential: citric acid is used in conjunction with antioxidants such as ascorbic and erythorbic acids, to inhibit colour and flavour deterioration caused by metal-catalysed enzymatic oxidation. Recommended usage levels are generally 0.1% to 0.3% with the antioxidant at 100 to 200 ppm.
d. Inactivate undesirable enzymes: oxidative browning in most fruits and vegetables is catalysed by the naturally present polyphenol oxidase. The enzymatic activity is strongly dependent on pH.

Addition of citric acid to reduce pH below 3 will result in inactivation of this enzyme and prevention of browning reactions.

Cherry and Apricot Oils are Safe for Food Use

The oils obtained by cold pressing the kernels of the cherry (Prunus cerasus) and the apricot (Prunus armeniaca) have been declared acceptable for food use by the UK Ministry of Agriculture, Fisheries &

Food's Advisory Committee on Novel Foods and Processes (ACNFP) by June 1993. In its assessments of the safety in use of the cherry and apricot kernel oils, the ACNFP consider specifications that included data on fatty acid composition, the presence of natural antioxidants and the content of cyanide, mycotoxins and heavy metals.

The Committee says that it gave particular consideration to the possible presence in the oils of the cyanogenic glucoside amygdalin, from which cyanide is released by enzymic action when the kernels of cherry and apricot are crushed. Amygdalin was found to be absent from the cherry and apricot kernel oils.

The ACNFP is satisfied that there are no food safety reasons why the use of cherry and apricot kernel should not be acceptable provided there is compliance with the specifications shown in Table below.

Table: *Specification of purity for cherry and apricot kernel oils as determined by UK ACNFP*

	Cherry	***Apricot***
Contaminants limits:		
Heavy metals (total)	0.5 mg/kg	0.5 mg/kg
Aflatoxins (total)	4.0 g/kg	4.0 g/kg
Cyanide	0.15 mg/kg	0.15 mg/kg
Pesticide residues	0.01 mg/kg	0.01 mg/kg
Tocopherols:		
alpha/delta/gamma (mg/kg)	356-886	569-899

Source: Anon. (1993)

The oils are obtained by the mechanical mincing and cold pressing of kernels extracted from cleaned cherry or apricot stones. After filtering, the oils are stored and are to be sold in a raw, unrefined state. The cherry and apricot kernel oils are high unsaturated and are expected to be used as speciality oils for salad dressings, baking and shallow frying applications.

The Use of Fruit Juices in Confectionery Products

During the last decade, the concept of fruit juices has gained immensely on consumer popularity. The majority of new non-alcoholic and alcoholic fruit drink products were a combination of syrups, fruit juices and flavours.

The confectionery industry followed suit and new products incorporated fruit juices as part of their confectionery formulations and processes. Fruit juice concentrates of high solids are often used instead of normal or single-fold juices. Juice concentrates are made of pure fruit juices. The process starts with pressing fruits and obtaining pure fruit juice; this is stabilised by heat treatment which inactivates enzymes and micro-organisms. The next processing step is concentration under vacuum up to 40-65° Brix or 4-7 fold. The concentrates are then blended for standardisation and stored.

These fruit juice concentrates are often further stabilised by the addition of sodium benzoate and potassium sorbate and are usually stored away from light and are refrigerated or frozen.

Depectinised fruit juices are also used to prevent foaming in confectionery processes and are essential for use in clear beverage products. Fruit juice concentrates which are depectinised, and have added preservatives are called stabilised, clarified, fruit juice concentrates. Fruit juices are used in confectionery products in conjunction with natural and artificial flavours which provides intense flavour impact and are cost-effective for a confectionery product.

The traditional concern in using fruit juice concentrates in confectionery applications has been the effect of the natural acids on the finished product, particularly the formation of invert sugar during processing. This is a logical concern since concentrates contain differing amounts and types of acids. For example: apple, cherry, strawberry and other berries contain primarily malic acid. Grapes mainly contain tartaric acid. Cranberry is high in quinic acid. Citrus fruits and pineapple contain differing amounts of citric acid. The concentrates, when used, are normally buffered to a pH of 5-7 with sodium hydroxide.

In formulating products with fruit juice concentrates, the solids of the concentrate are considered as mostly reducing sugars and a reduction in corn syrup is made to compensate for equivalent amount of reducing sugar being added in the concentrate.

The exact replacement can be determined by measuring the D.E. of the concentrate to be added. In formulations when small amounts of concentrate are used (less than 1%), no adjustment is made since the reducing sugar contribution of the concentrate is not significant. Fruit juice concentrates can also be used to provide a source of natural colour, in particular red colour. Grape, raspberry, cherry, strawberry and cranberry concentrates in small amounts are very effective in colouring cream centres.

The inclusion of fruit juices in confectionery products is now left up to the imagination of the manufacturer. These products must, of course, hold up to the standards of flavour integrity, and product excellence, during the shelf-life of these products.

Vegetable Specific Processing Technologies

Vegetable processors must appreciate the substantial differences that varieties of a given vegetable will possess. In addition to variety and genetic strain differences with respect to weather, insect and disease resistance, varieties of a given vegetable will differ in size, shape, time of maturity, and resistance to physical damage.

Varietal differences then further extend into warehouse storage stability, and suitability for such processing methods as canning, freezing, pickling or drying. A variety of peas that is suitable for canning may be quite unsatisfactory for freezing and varieties of potatoes that are preferred for freezing may be less satisfactory for drying or potato chip manufacture. This should be expected since different varieties of a given vegetable will vary somewhat in chemical composition, cellular structure and biological activity of their enzyme system.

Harvesting and Pre-processing

When vegetables are maturing in the field they are changing from day to day. There is a time when the vegetable will be at peak quality from the stand-point of colour, texture and flavour.

This peak quality is quick in passing and may last only a day. Harvesting and processing of several vegetables, including tomatoes, corn and peas are rigidly scheduled to capture this peak quality.

After the vegetable is harvested it may quickly pass beyond the peak quality condition. This is independent of microbiological spoilage; these main deteriorations are related to:

a) loss of sugars due to their consumption during respiration or their conversion to starch; losses are slower under refrigeration but there is still a great change in vegetable sweetness and freshness of flavour within 2 or 3 days;

b) production of heat when large stockpiles of vegetables are transported or held prior to processing.

   At room temperature some vegetables will liberate heat at a rate of 127,000 kJ/ton/day; this is enough for each ton of vegetables to melt 363 kg of ice per day. Since the heat further deteriorates the vegetables and speeds micro-organisms growth, the harvested vegetables must be cooled if not processed immediately.

   But cooling only slows down the rate of deterioration, it does not prevent it, and vegetables differ in their resistance to cold storage. Each vegetable has its optimum cold storage temperature which may be between about 0-100 C (32-500 F).

c) the continual loss of water by harvested vegetables due to transpiration, respiration and physical drying of cut surfaces results in wilting of leafy vegetables, loss of plumpness of fleshy vegetables and loss of weight of both.

Moisture loss cannot be completely and effectively prevented by hermetic packaging. This was tried with plastic bags for fresh vegetables in supermarkets but the bags became moisture fogged, and deterioration of certain vegetables was accelerated because of buildup of CO2 and decrease of oxygen in the package. It therefore is common to perforate such bags to prevent these defects as well as to minimise high humidity in the package which would encourage microbial growth.

Shippers of fresh vegetables and vegetable processors, whether they can, freeze, dehydrate, or manufacture soups or ketchup, appreciate the instability and perishability of vegetables and so do everything they can to minimise delays in processing of the fresh product. In many processing plants it is common practice to process vegetables immediately as they are received from the field.

To ensure a steady supply of top quality produce during the harvesting period the large food processors will employ trained field men; they will advise on growing practices and on spacing of plantings so that vegetables will mature and can be harvested in rhythm with the processing plant capabilities. This minimises stockpiling and need for storage.

Cooling of vegetables in the field is common practice in some areas. Liquid nitrogen-cooled trucks may next provide transportation of fresh produce to the processing plant or directly to market.

Upon arrival of vegetables at the processing centre the usual operations of cleaning, grading, peeling, cutting and the like are performed using a moderate amount of equipment but a good deal of hand labour also still remains.

Reception. This covers qualitative and quantitative control of delivered vegetables. The organoleptic control and the evaluation of the sanitary state, even if they are very important steps in vegetables' characteristics assessment, cannot establish their technological value.

On the other hand, laboratory controls do not precisely establish their technological properties because of the difficulty in putting into showing some deterioration when using rapid control methods.

One correct method of vegetable quality appraisal is their overall evaluation based on the whole complex of data that can be obtained by combining an extensive organoleptic evaluation with simple analysis that can be performed rapidly in plant laboratory. These analysis can be:

a. refractometric extract (tomatoes, fruit, etc.);
b. specific weight (potatoes, peas, etc.);
c. consistency (measured with tenderometers, penetrometers, etc.);
d. boiling tests, etc.

Temporary storage. This step should be as short as possible and better completely eliminated. Vegetables can be stored in:

a. simple stores, without artificial cooling;
b. in refrigerated stores; or, in some cases,
c. in silos (potatoes, etc.).

Simple stores should be covered, fairly cool, dry and well ventilated but without forced air circulation which can induce significant losses in weight through intensive water evaporation; air relative humidity should be at about 70-80%. Refrigerated storage is always preferable and in all cases a processing centre needs a cold room for this purpose, adapted in volume I capacity to the types and quantities of vegetables (and fruits) that are further processed.

Washing. Washing is used not only to remove field soil and surface micro-organisms but also to remove fungicides, insecticides and other pesticides, since there are laws specifying maximum levels of these materials that may be retained on the vegetable; and in most cases the allowable residual level is virtually zero. Washing water contains detergents or other sanitisers that can essentially completely remove these residues.

The washing equipment, like all equipment subsequently used, will depend upon the size, shape and fragility of the particular kind of vegetable:

- flotation cleaner for peas and other small vegetables;
- rotary washer in which vegetables are tumbled while they are sprayed with jets of water; this type of washer should not be used to clean fragile vegetables;

Sorting. This step covers two separate operations:

a) removal of non-standard vegetables (and fruit) and possible foreign bodies remaining after washing;
b) quality grading based on variety, dimensional, organoleptical and maturity stage criterion.

Skin Removal/peeling. Some vegetables require skin removal. This can be done in various ways.

Mechanical : This type of operation is performed with various types of equipment which depend upon the result expected and the characteristics of the fruit and vegetables, for example:

i. a machine with abrasion device (potatoes, root vegetables);
ii. equipment with knives (apples, pears, potatoes, etc.);
iii. equipment with rotating sieve drums (root vegetables). Sometimes this operation is simultaneous with washing (potatoes, carrots, etc.) or preceded by blanching (carrots).

Chemical: Skins can be softened from the underlying tissues by submerging vegetables in hot alkali solution. Lye may be used at a concentration of about 0.5-3%, at about 93° C (2000 F) for a short time period (0.5-3 min). The vegetables with loosened skins are then conveyed under high velocity jets of water which wash away the skins and residual lye.

In order to avoid enzymatic browning, this chemical peeling is followed by a short boiling in water or an immersion in diluted citric acid solutions.

It is more difficult to peel potatoes with this method because it is necessary to dissolve the cutin and this requires more concentrated lye solutions, up to 10%.

Thermal: Wet heat (steam). Other vegetables with thick skins such as beets, potatoes, carrots and sweet potatoes may be peeled with steam under pressure (about 10 at) as they pass through cylindrical vessels.

This softens the skin and the underlying tissue. When the pressure is suddenly released, steam under the skin expands and causes the skin to puff and crack. The skins are then washed away with jets of water at high pressure (up to 12 at).

Dry heat (flame). Other vegetables such as onions and peppers are best skinned by exposing them to direct flame (about 1 min at 1000° C) or to hot gases in rotary tube flame peelers. Here too, heat causes steam to develop under skins and puff them so that they can be washed away with water.

Manual peeling only use when the other methods are impossible or sometimes as a completion of the other three ways. Average losses at this step are given in Table below.

Table: *Losses at vegetable peeling, in %*

	Peeling methods		
Vegetables	***Manual***	***Mechanical***	***Chemical***
Potatoes	15-19	18-28	-
Carrots	13-15	16-18	8-10
Beets	1416	13-15	9-10

Size reduction. This step is applied according to specific vegetable and processing technology requirements.

Blanching. The special heat treatment to inactivate enzymes is known as blanching. Blanching is not indiscriminate heating. Too little is ineffective, and too much damages the vegetables by excessive cooking, especially where the fresh character of the vegetable is subsequently to be preserved by processing. This heat treatment is applied according to and depends upon the specificity of vegetables, the objectives that are followed and the subsequent processing/ preservation methods.

Two of the more heat resistant enzymes important in vegetables are catalase and peroxidase. If these are destroyed then the other significant enzymes in vegetables also will have been inactivated. The heat treatment to destroy catalase and peroxidase in different vegetables are known, and sensitive chemical tests have been developed to detect the amounts of these enzymes that might survive a blanching treatment.

Because various types of vegetables differ in size, shape, heat conductivity, and the natural levels of their enzymes, blanching treatments have to be established on an experimental basis. As with sterilisation of foods in cans, the larger the food item the longer it takes for heat to reach the centre. Small vegetables may be adequately blanched in boiling water in a minute or two, large vegetables may require several minutes.

Blanching as a unit operation is a short time heating in water at temperatures of 100° C or below. Water blanching may be performed in double bottom kettles, in special baths with conveyor belts or in modern continuous blanching equipment.

In order to reduce losses of hydrosoluble substances (mineral salts, vitamins, sugars, etc.) occurring during water blanching, several methods have been developed:

- temperature setting at 85-95° C instead of 100° C;
- blanching time has to be just sufficient to inactivate enzymes catalase and peroxidase;
- assure elimination of air from tissues.

An illustration of blanching parameters is seen in Table below.

Table: *Blanching parameters for some vegetables*

Vegetables	***Temperature, °C***	***Time, min.***
Peas	85-90	2-7
Green beans	90-95	2-5
Cauliflower	Boiling	2
Carrots	90	3-5
Peppers	90	3

Steam heat treatment can also be applied instead of water blanching as a preliminary step before freezing or drying, as long as the preservation method is only used for enzyme inactivation and not to modify consistency. For drying, the vegetables are conveyed directly from steaming equipment to drying installations without cooling. Vegetable steaming is carried out in continuous installations with conveyer belts made from metallic sieves.

Cooling of vegetables after water blanching or steaming is performed in order to avoid excessive softening of the tissues and has to follow immediately after these operations; one exception is the case of vegetables for drying which can be transferred directly to drying equipment without cooling.

Natural cooling is not recommended because is too long and generates significant losses in vitamin C content. Cooling in pre-cooled air (from special installations) is sometimes used for vegetables that will be frozen.

Cooling in water can be achieved by sprays or by immersion; in any case the vegetables have to reach a temperature value under 37° C as soon as possible. Too long a cooling time generates supplementary losses in valuable hydrosoluble substances; in order to avoid this, the temperature of the cooling water has to be as low as possible.

Canning: Large quantities of vegetable products are canned. A typical flow sheet for a vegetable canning operation (which also applies to fruit for the most part) covers some food process unit operations performed in sequence: harvesting; receiving; washing; grading; heat blanching; peeling and coring; can filling; removal of air under vacuum;

sealing/closing, retorting/heat treatment; cooling; labelling and packing. The vegetable may be canned whole, diced, puréed, as juice and so on.

On-line Simplified Methods for Enzyme Activity Check

Peroxidase test:

Solutions. In order to check the peroxidase activity two solutions have to be prepared:

- 1% guaiacol in alcohol solution (1 g guaiacol is dissolved in about 50 cm^3 of 96% ethylic alcohol and then this preparation is brought to 100 cm° with the same solvent);
- peroxide solution 0.3% (1 cm^3 perhydrol is brought to 100 cm^3 with distilled water.

Sampling. From various parts of the material samples are taken (about 20-30 pieces, etc.); the material is then crushed in a laboratory bowl in order to obtain an average sample.

Check. Prom the average sample, 10-20 g of material is taken in a medium capacity test tube; on this sample are poured: 20 cm^3 distilled water; 1 cm^3 of 1% guaiacol solution; 1.6 cm^3 of peroxide solution. The contents of the test tube is shaken well. The gradual appearance of a weak pink colour indicates an incomplete peroxidase inactivation - reaction slightly positive. If there are no tissue colour modifications after 5 minutes, the reaction is negative and the enzymes have been inactivated. As an orientative check it is also possible to simply pour a few drops of 1% guaiacol solution and 0.3% peroxide solution directly on blanched and crushed vegetables. A rapid and intensive brown-reddish tissue colouring indicates a high peroxidase activity (positive reaction).

Catalase Test

In order to identify the catalase enzyme activity, 2 g of dehydrated vegetables are well crushed and mixed with about 20 cm^3 of distilled water. After 15 min softening, 0.5 cm^3 of a 0.5% or 1% peroxide solution is poured on prepared vegetables. In the presence of catalase, a strong oxygen generation is observed for about 2-3 minutes. These tests are of a paramount importance in order to determine the vegetable blanching treatments (temperature and time); incomplete enzyme inactivation has a negative effect on finished product quality. For cabbage catalase inactivation by blanching is sufficient; blanching further to peroxidase inactivation would have negative effects on product quality and even complete browning.

For all other vegetables and for potatoes, both tests MUST be negative, for catalase and for peroxidase.

Quality control/quality assurance and international trade; good manufacturing practices (gmp); hygiene requirements; hazard analysis and critical control points (haccp).

Quality Control/Quality Assurance and International Trade

The international trade in processed fruits and vegetables is very large with an ever increasing number of different types being processed and exported. Whereas once, processing was limited to mostly temperate climate fruits and vegetables, the change has now broadened to include tropical and subtropical types. The reasons are twofold. Firstly, consumers' dietary habits have become more diverse so that, for example people living in North America may very well like fruit and vegetables grown in Africa or Asia. Secondly, processing techniques, whether they be for canning, freezing or drying, have been improved to an extent where final product is palatable, nutritious and of long and reliable shelf life.

Many developing countries have taken advantage of the continuing worldwide demand for processed fruits and vegetables and earned valuable foreign exchange from exports of products to profitable markets. The export quality control and inspection of processed fruits and vegetables is directed at ensuring that the final products: have been processed in a registered export establishment that is constructed, equipped and operated in an hygienic and efficient manner; conform to the requirements of the export regulations for processed fruits and vegetables, and those of the importing country, in respect of such things as quality grades, defects, ingredients, packaging materials, styles, additives, contaminants, fill of container, drained weight; and, conform to labelling requirements.

Inspection and Certification Procedures

In most countries, in processing fruits and vegetables for export, it is not customary to apply continuous inspection as it is in the case of meat. Few, if any, importing countries require it, and the nature of the products themselves is such that only part time check inspection is required during processing together with statistically based inspection, including sampling and analysis, of final product.

However, in circumstances where an establishment is processing export product for the first time, it can be argued that there is an advantage in adopting continuous inspection until the operation is satisfactorily established.

In any event, inspection of raw materials should be carried out at the commencement of each processing run to ensure that only sound fruit or vegetables of sufficient maturity (degree of ripeness) are used for processing. Sampling checks of raw materials should be carried out as frequently as the inspector thinks necessary. The inspector must ensure that adequate hygiene practices are followed during the processing of the product. For example, in the case of canned and frozen products and other processing methods, raw materials should be washed absolutely clean so that fruit and vegetables entering the processing line are free from dirt, superficial residues of agricultural chemicals, insects and extraneous plant material.

In the case of dried product, especially where the raw material is sun-dried on drying greens or racks, care must be taken to minimize contamination by bird and animal droppings, dust and extraneous plant material. It is often necessary to wash the dried product to ensure cleanliness of the final product.

In the case of canning and freezing, the inspector must obtain full details of the processing programme for at least the following day from management, so that an adequate inspection programme can be scheduled. In much the same way as for fresh fruit and vegetables, the inspector must also be aware of the pesticides and other chemicals used in the production of the raw materials. Necessary laboratory analyses can then be arranged to ensure residue levels in the final product do not exceed tolerances adopted by importing countries.

At the commencement of and during processing, the inspector should pay attention to the state of raw materials, the preparation of raw materials for processing (peeling, slicing, dicing, blanching, etc.), preparation and density of packing medium (sugar syrup, salt brine, etc.), the state of cans or containers to be used (cleanliness and strength), the cooking or freezing process (time/temperature relationship), can filling and closure and can/container storage.

After processing, the inspector should check the final product to ensure the drained or thawed weight, the vacuum and headspace, packing medium strength and that can/container conditions are satisfactory. Statistically based sampling plans should be adopted for the examination of final product to ensure it meets the requirements of the export regulations. The labelling applied to cans/containers should also be checked to ensure both their correctness and compliance with the export regulations and the requirements of those countries in which the product is to be marketed.

Cans should also be examined to make sure that the correct embossing relating to the product, its date of production and the registered number of the export establishment has been applied. Each establishment registered for the export of processed fruit and vegetables or for canned or frozen foods should have its own quality laboratory sufficiently equipped and staffed to carry out physical, chemical and microbiological examinations of the goods.

Inspectors should have access to the laboratory facilities and the establishment's quality control records as and when required. Independent laboratory examination of product should be made by the agency having responsibility for export on the basis of a statistically developed sampling plan. In those countries where fruit and vegetable production is a seasonal event, processing for export generally takes place at the time of peak production and then declines, often to a halt, as the supply of raw materials declines. As a result, most export establishments produce at their peak of production far more product than they export at that time.

Therefore, most manufacturers find it necessary to store product for considerable periods before it is exported.

Thus, proper storage is essential if the product is to retain its quality and cans remain untarnished. Inspectors should regularly inspect storage facilities, noting their conditions and that of the stored product, looking for signs of deterioration such as pest infestation and rusting of cans.

Prior to export, the exporter should be required to notify the export quality and inspection agency of his intention to export in accordance with the provisions of the export processed fruits and vegetables regulations and on the prescribed "Notice of Intention to Export" form.

The notice should be submitted in sufficient time before the shipment date to enable the product to be inspected satisfactorily; the intensity of inspection depending on the original state of the product, the conditions under which it has been stored and the length of storage. When product is approved, the agency will issue the exporter an "Export Permit" authorizing Custom's clearance of the product.

Export Quality Control and Inspection Systems for Foods

With the advent and development of a food consciousness amongst consumers, stimulated by the work of the Joint FAO/WHO Codex Alimentarius Commission through its elaboration of food standards, codes of hygienic practice and the Code of Ethics for International Trade in Food, an increasing number of countries have adopted sophisticated food laws and established food control agencies, some with the aid of FAO. Consequently, those countries no longer accept products on trust that they are satisfactory, but instead, demand that food imports meet the requirements of their food laws and pass inspection by their control agencies. Moreover, many of them require exporting countries to certify that products comply with their national legislation and some also require additional special declarations.

As a result of these developments the emphasis of activity of Export Quality Control and Inspection Systems has changed. Although most of them still establish their own standards of quality control and adopt standards for foods for export, most of their effort and resources are now directed at ensuring that foods for export meet the mandatory requirements of importing countries and providing the necessary associated certification. To do otherwise is to invite either the detention or, at worst, rejection of product at point of entry.

Detentions and Rejections

Food exporting countries can no longer assume that there is a good chance that products not complying with the requirements of importing countries will escape the inspection at the point of entry.

Details of foods imports released by the United States Food and Drug Administration (FDA) indicate that significant quantities of product are at least detained, and at worst rejected, because they fail to meet U.S. food laws.

Reasons given for the detentions include:

- non compliance with labelling requirements;
- decomposition;
- insect and animal filth and damage;
- use of prohibited additives;
- non compliance with requirements of the U.S. low acid canned food regulations;
- heavy metal contamination;
- excessive levels of pesticide residues;
- excessive levels of mycotoxin;
- mould infestation;
- microbiological contamination;
- swollen and otherwise faulty cans.

The message for food exporting countries is quite clear - ensure your products comply with the mandatory requirements of importing countries or run the very real risk of having them rejected at considerable financial loss to the exporter and the country and resulting in damage to the commercial reputation of both.

While the foregoing relates to the U.S.A. experience, because it is the only country that currently publishes data about detentions and rejections of food imports it can be assumed that record more or less reflects the experience of other food importing countries. It might well be asked why such significantly high levels of detentions and rejections of food imports take place. Undoubtedly the reasons are many and varied. However, the evidence shows that the most important reasons include:

- the inability of some export food industries, especially in developing countries, to handle, process, package and transport products to meet the mandatory requirements of importing countries;
- lack of awareness by food exporting countries of the mandatory requirements of importing countries, including certification;
- lack of adequate export control programmes and related agencies in food exporting countries, preventing them from exercising the necessary product surveillance and giving reliable and credible certification, and
- a lack of communication, between food control authorities and agencies in exporting and importing countries.

All four can be remedied by governments if they possess sufficient political will and take the necessary steps to do so.

Good Manufacturing Practices (GMP); Hygiene Requirements

Any person who has an illness, open lesions, including boils, sores, infected wounds, or any other abnormal source of microbial contamination must not work in any operation (in a food processing centre) which could result in the food, food-contact surface, or food packaging materials becoming contaminated.

Cleanliness

The following applies to people who work in direct contact with food preparation, food ingredients or surfaces of equipment or utensils that will contact food: they must wear clean outer garments, maintain a high degree of personal cleanliness and conform to hygienic practices while on duty; they must wash their hands thoroughly and, if they are working at a job where it is necessary, they must also sanitize their hands before starting work, after each absence from the workstation and at any other time when the hands have become soiled or contaminated; they must also remove all unsecured jewellery. People who are actually handling food, should remove any jewellery that cannot be properly sanitized from their hands; it is necessary to wear effective hair restraints, such as hairnets, caps, headbands or beard covers; operators must not store clothing or other personal belongings in food processing areas. Also, eating food, drinking beverage or using tobacco (in any form) must not be allowed in food processing area; all necessary steps have to be taken by supervisors to prevent operators from contaminating foods with microorganisms or foreign substances such as perspiration, hair, cosmetics, tobacco, chemicals and medicants.

Education and Training

Persons who are monitoring the sanitation programs must have the education and/or experience to demonstrate that they are qualified. Food handlers and supervisors should receive training that will make them aware of the danger of poor personal hygiene and unsanitary work habits.

Supervision

Someone must be assigned the responsibility that all personnel will comply with all the requirements of these GMP's.

Plants and Grounds

Grounds around a food processing centre which are under the control of this centre must be free from conditions such as: improperly stored equipment; litter, waste or refuse; uncut weeds or grass close to buildings; excessively dusty roads, yards or parking lots; inadequately drained areas - potential foot-borne filth or breeding places for insects or microorganisms; inadequately operated systems for waste treatment and disposal. Plant construction and design shall: provide enough space for sanitary arrangement of equipment and storage of materials; floors, walls and ceilings must be constructed so that they are cleanable and must be kept clean and in good repair; separate by partition, location, time and other means, any operations that may cause cross-contamination of food products with undesirable microorganisms, chemicals, filth or other extraneous material; provide effective screening or other protection to keep out birds, animals and vermin such as insects and rodents. Provide adequate ventilation to prevent contamination of foods with odours, noxious fumes or vapours (including steam); light bulbs, skylights or any other glass must be of the safety type or protected so that glass contamination cannot occur in case of breakage.

Sanitary Operations

General maintenance. The plant and all fixtures must be kept in good repair and be maintained in a sanitary condition. Cleaning operations must be conducted in a manner that will minimize the possibility of contaminating foods or equipment surfaces that contact food.

Pest control:

- No animals or birds are allowed anywhere in the plant
- Programs must be in effect to prevent contamination by animals, birds and pests, such as rodents and insects;

- Insecticides and rodenticides may be used as long as they are used properly (according to label instructions);
- These pesticides must not contaminate food or packaging materials with illegal residues;

Sanitation of equipment and utensils:

- Utensils and equipment surfaces that are in contact with food must be cleaned as often as necessary to prevent food contamination;
- Equipment surfaces that are not in contact with food should be cleaned as frequently as necessary to minimize accumulation of dust, dirt, food particles, etc.
- Single-service articles such as disposable utensils, paper cups, paper towels, etc., should be:
- Stored in appropriate containers;
- Handled, dispensed, used and disposed of in a manner that prevents contamination of food or equipment;
- Where there is the possibility of introducing undesirable microorganisms into food, all utensils and equipment surfaces that contact food must be cleaned and sanitized before use and following any interruption during which they may have become contaminated;
- When utensils or equipment are used in a continuous production operation, they must be cleaned and sanitized on a predetermined schedule;
- Any facility, procedure, machine or device may be used for cleaning and sanitizing, as long as it has been established that the procedure will do the job effectively.

Storage and handling of clean portable equipment and utensils:

a) This refers to portable equipment or utensils which have surfaces that will contact foods;

b) When such equipment or utensils have been cleaned and sanitized, they should be stored in a manner that will protect the food contact surfaces from splash, dust and other contamination.

Sanitary Facilities and Controls

Water supply. Any water that comes into contact food or processing equipment must be safe and of adequate sanitary quality.

Sewage disposal. Must flow into an adequate sewage system or disposed of through other adequate means.

Plumbing. Must be of adequate size and design to:

a. Supply enough water to areas in the plant where it is needed;
b. Properly convey sewage or disposable liquid waste from the plant;
c. Not create a source of contamination or unsanitary condition;
d. Provide adequate floor drainage where hosing-type cleaning is done or where operations discharge water or liquid waste onto the floor;
e. Insure that there is no backflow from cross-connection between piping systems that discharge waste water or sewage, and those that carry water for food or food manufacturing.

Toilet facilities:

a. Toilets and hand-washing facilities must be provided inside the fruit and vegetable processing centres;
b. Toilet tissue must be provided;
c. Toilets must be kept sanitary and in good repair;
d. Toilet rooms must have self-closing doors;
e. Toilet rooms must not open directly into areas where food is exposed unless steps have been taken to prevent airborne contamination (example: double doors, positive airflow, etc.);
f. Signs must be posted that direct employees to wash their hands with soap or detergent after using the toilet.

Hand-washing Facilities

Adequate and convenient hand-washing and, if necessary, hand-sanitizing facilities must be provided anywhere in the plant where the nature of employees jobs requires that they wash, sanitize and dry their hands;

These hand-washing facilities must provide:

- Running water at a suitable temperature;
- Effective hand-cleaning and hand-sanitizing preparations;
- Clean towel service or suitable drying devices;
- Easily cleanable waste receptacle;
- Water control valves designed and constructed to protect against recontamination of clean, sanitized hands;
- Signs directing employees handling unprotected food to wash and, if appropriate, sanitize theirs hands before starting work, after each absence from the workstation, and any other time when the hands have become soiled or contaminated.

Rubbish and offal disposal must be handled in such a manner that they do not serve to attract or harbour pests or create contaminating conditions.

Equipment and utensils:

a. Equipment and utensils must be designed and constructed so that they are adequately cleanable and will not adulterate food with lubricants, fuel, metal fragments, contaminated water, etc.
b. Equipment should be installed so that it, and the area around it, can be cleaned;
c. Food contact surfaces shall be made of nontoxic materials and must be corrosion-resistant;
d. Seams on food contact surfaces shall be smoothly bonded, or maintained in order to minimize the accumulation of food particles, dirt and organic matter;
e. Equipment in processing areas that does not come into contact with food shall be constructed so that it can be kept clean;
f. Holding, conveying and manufacturing systems, including gravimetric, pneumatic, closed and automated systems, shall be maintained in a sanitary condition;
g. Each freezer and cold storage compartment shall have an indicating thermometer, temperature measuring or recording device, and should have an automatic control for regulating temperature, or an automatic alarm system to indicate a significant temperature change;
h. Instruments and controls used for measuring, regulating or recording temperatures, pH, acidity, water activity, etc. shall be adequate in number, accurate and maintained.

Processes and Controls

There must be an individual who is responsible for supervising the overall sanitation of the plant.

Raw materials and ingredients:

a. Must be inspected and sorted to insure that they are clean, wholesome and fit for processing into human food;
b. Must be stored under conditions that will protect against contamination and minimize deterioration;
c. Must be washed or cleaned to remove soil and other contamination:
   - Water used for washing, rising or conveying food products must be of sanitary quality;

- Water must not be reused for washing, rinsing or conveying if contamination of food may result;
- Containers and carriers (such as trucks or railcars) should be inspected to assure that their condition has not contaminated raw ingredients;

d. Raw materials shall not contain levels of microorganisms that may produce food poisoning or other disease, or they shall be pasteurized or otherwise treated during manufacturing operations so that the product will not be adulterated;

e. Materials susceptible to contamination with natural toxins, e.g., aflatoxin, shall comply with national and international official levels before they are incorporated into the finished food;

f. Materials susceptible to contamination with pests, undesirable microorganisms, or extraneous material, shall comply with national and international regulations, guidelines and defect action levels;

g. Materials shall be stored in containers, and under conditions which protect against contamination;

h. Frozen materials shall be kept frozen. If thawing is required prior to use, it shall be done in a manner that prevents contamination.

Manufacturing Operations

a. Food processing equipment must be keep in a sanitary condition through frequent cleaning and, when necessary, sanitizing. If necessary, such equipment must be taken apart for thorough cleaning.

b. It is necessary to process, package and store food under conditions that will minimize the potential for undesirable microbiological growth, toxin formation, deterioration or contamination. To accomplish this may require careful monitoring of such factors as time, temperature, humidity, pressure, flow rate, etc. The object is to assure that mechanical breakdowns, time delays, temperature fluctuations or other factors do not allow the foods to decompose or become contaminated.

c. Food shall be held under conditions that prevent the growth of undesirable microorganisms as follows:

- Refrigerated foods shall be maintained at 45° F or below;
- Frozen foods shall be maintained in a frozen state;

- Acid or acidified foods to be held in hermetically sealed containers at ambient temperatures shall be heat-treated to destroy mesophyllic microorganisms;

d. Measures such as sterilizing, irradiating, pasteurizing, etc., shall be adequate to destroy or prevent the growth of undesirable microorganisms;

e. Work-in-process shall be protected against contamination;

f. Finished food shall be protected from contamination;

g. Equipment, containers and utensils shall be constructed, handled and maintained to protect against contamination;

h. Measures, e.g., sieves, traps, metal detectors, shall be used to protect against the inclusion of metal or other extraneous material in food;

i. Food or materials that are adulterated shall be disposed of in a manner that prevents other food from being contaminated;

j. Mechanical manufacturing steps such as washing, peeling, etc., shall be performed to protect against contamination by providing adequate protection from contaminants that may drip, drain or be drawn into the food, by adequately cleaning and sanitizing all food-contact surfaces and by using time and temperature controls at and between each manufacturing step;

k. Heat-blanching should be done by heating the food to the required temperature, holding it at this temperature for the required time, and then either rapidly cooling the food or passing it to the next manufacturing step without delay;

l. Filling, assembling, packaging, and other operations shall be performed in such a way that the food is protected against contamination by:
   - Use of a quality control operation in which the Critical Control Points are identified and controlled during manufacturing;
   - Adequate cleaning and sanitizing of all food-contact surfaces and food containers;
   - Using materials for food containers and food-packaging materials that are safe and suitable;
   - Providing physical protection from contamination, particularly airborne contamination;
   - Using sanitary handling procedures.

m. Food such as, but not limited to, dry mixes, nuts, intermediate moisture food, and dehydrated food, that relies on the control of aw for preventing the growth of undesirable microorganisms shall be processed to and maintained at a safe moisture level by:
   - Monitoring the aw of food;
   - Controlling the soluble solids/water ratio in finished food;
   - Protecting finished food from moisture pick-up, by use of a moisture barrier, or by other means, so that the Aw of the food does not increase to an unsafe level;

n. Food such as, but not limited to, acid and acidified food, that relies principally on the control of pH for preventing the growth of undesirable microorganisms shall be monitored and maintained at a pH of 4.6 or below by: - Monitoring the pH of raw materials, food in process, and finished food; - Controlling the amount of acid or acidified food added to low-acid food;

o. If ice is used and comes in contact with food products, it must be made from potable water and be in a sanitary condition;

p. Areas and equipment that are used to process human food should not be used to process non-human food-grade animal feed, or inedible products unless there is no possibility of contaminating the human food;

q. A coding system should be utilized that will allow positive lot identification in the event it is necessary to identify and segregate lots of food that may be contaminated.
   - Records should be kept for a period of time that exceeds the self life of the product, except that
   - Records need not be kept beyond two years.

Chapter 7

Biotechnology of Vegetables and Fruits

The plant is the most important agent in crop production. Soils, cultivation, fertilizers, irrigation, and other factors, in a sense, are all more or less subsidiary. Soils are modified by cultivation, by adding manure or other fertilizers, by drainage or irrigation, and in other ways with the express purpose of changing the environment so as to stimulate plants to increased productivity. Hence, it is not surprising that from time immemorial extended observations and, later, experiments have been made upon the aerial growth of crops under varying conditions. In fact an almost bewildering array of literature has resulted. But quite the converse is true of the underground parts. The root development of vegetable crops has received relatively little attention, and indeed accurate information is rarely to be found. The roots of plants are the least known, least understood, and least appreciated part of the plant. This is undoubtedly due to the fact that they are effectually hidden from sight. Notwithstanding the extreme difficulty and tediousness of laying the roots bare for study, it is not only remarkable but also extremely unfortunate that such investigations have been so long neglected.

It clearly seems that a thorough understanding of the activities of plants both above ground and below ground and the ways in which these activities are favourably or unfavourably modified by various cultural practices should be basic for scientific crop production. Yet almost countless field experiments, selections, breeding, and testing of varieties, etc. have been carried on in all parts of the world with little or no knowledge as to the behaviour of that very essential portion of the plant, the absorbing system. Similarly, in the study of soils, the greatest attention is given to the problems of the physics, chemistry, and bacteriology of this substratum and rather largely to the cultivated portion of the surface only. The soil, barring the living organisms which it supports, is perhaps the most complex, the most interesting, and the most wonderful thing in nature. Surely it should receive thorough investigation. But a study of the soil and the way in which its various relations affect yield without a consideration of the essential, intermediary absorbing system is more or less empirical. A complete, scientific understanding of the soils-crops relations cannot be attained until the mechanism by which the soil and the plant are brought into favourable relationships, i.e., the root system, is also understood.

The student of plant production should have a vivid, mental picture of the plant as a whole. It is just as much of a biological unit as is an animal. The animal is visible as an entity and behaves as one. If any part is injured, reactions and disturbance of the whole organism are expected. But in the plant, our mental conception is blurred by the fact that one of the most important structures is underground. Nor is the plant usually treated as an entity; it is often mutilated by pruning, cutting, and injuring the root system, frequently without much regard to the effect upon the remaining portion.

Modifying the Root Environment

In both field and garden the part of the plant environment that lies beneath the surface of the soil is more under the control of the plant grower than is the part which lies above. He can do relatively little toward changing the composition, temperature, or humidity of the air, or the amount of light. But much may be done by proper cultivation, fertilizing, irrigating, draining, etc. to influence the structure, fertility, aeration, and temperature of the soil. Thus, a thorough understanding of the roots of plants and the ways in which they are affected by the properties of the soil in which they grow is of the utmost practical importance. Before other than an empirical answer can be given to the questions as to what are the best methods of preparing the land for any crop, the type of cultivation to be employed, the best time or method of applying fertilizers, the application and amount of water of irrigation, kind of crop rotations, and many other questions, something must be known of the character and activities of the roots that absorb water and nutrients for the plant and the position that they occupy in the soil.

Adaptation of Roots to Environment

By more or less profound modifications of their root system, many plants become adapted to different soil environments, others are much less susceptible to change. Among forest trees, for example, the initial or juvenile root system of each species follows a fixed course of development and maintains a characteristic form for a rather definite period of time following germination. The tendency to change when subjected to different external conditions becomes more pronounced as the seedlings become older. But some species thus subjected exhibit much earlier tendencies to change than others and a widely different degree of flexibility is also shown. Hence, certain species, *e.g.*, red maple (*Acer rubrum*), because of the great plasticity of their root systems, are able to survive, at least for a time, in various situations from swamps to dry uplands. The roots of others such as bald cypress (*Taxodium distichum*) are so inflexible that they can grow only under certain favourable conditions and their distribution is thus greatly limited. Great variability also occurs in the rooting habits of fruit trees.

Native herbaceous species usually show great plasticity of root habit, successfully adapting themselves to considerable differences in soil environment. A few seem to be quite fixed in their habit of root growth.

Continued study has shown that many field crops, although governed first of all by the hereditary growth characters of the species or variety, are usually subject to change. Certain varieties are able to adapt their root systems to unfavourable conditions much more readily than others. The wide range in soil and agricultural conditions under which vegetable crops are grown renders them particularly suitable both for the investigations of their root habits and also for a study of the agricultural significance of the differences encountered.

Root Adaptation and Crop Production

Enough work has been done to show clearly that among garden crops root adaptations frequently occur. These will be pointed out in the following s, although a mere beginning has been made. The field is enormously large and difficult but one rich in possibilities. It seems entirely probable that some of the best-yielding crops are able to outstrip others largely because of their greater efficiency in securing a greater and more constant supply of water and nutrients. On the other hand, the failure of a crop to thrive in a particular soil may be due to a lack of adaptation of its root system to the environment imposed upon it. Both of these conditions are illustrated by the growth of flax in India.

The agriculture in many of these areas is ancient, there have been few innovations, and the soil conditions have had time to impress themselves on the varieties of crops cultivated. A condition of equilibrium between the type of plant and the soil has been obtained, as there has been ample time for the operation of natural selection.

When we compare the root system of linseed from the black-soil areas with that of the varieties grown on the Gangetic alluvium, striking differences appear. The roots produced on the black soils (of the Peninsula) are deep, somewhat sparse, and are well adapted to ripen the plant quickly with the minimum of moisture. The type of gearing fits the soil. On the alluvium, where moisture is more abundant and where the aeration of the subsoil is poor, the root system is superficial but at the same time well developed. On the intermediate types of soil... linseed produces a type of root about halfway between that of the black soils and of the alluvium. Once more the root system is found to fit the soil type.

Further when we grow side by side on the alluvium these three classes of linseed, there is little or no adaptation of the roots to the new conditions, but the three types behave very much as they would in their native habitat. The deep, sparse root system of the black soil areas is developed in the alluvium, although it is fatal to the well-being of the crop. When, the experiment is reversed and the types which suit the alluvium are grown on the black soils, there is again little or no adaptation to fit the new conditions. The linseed crop consists of a large number of varieties which differ from one another in all sorts of characters, including the extent and distribution of the roots. The root systems of the varieties are just as characteristic and just as fixed as the differences in the seed and other above ground characters of these plants. A similar state of affairs obtains in other crops like wheat... and the opium poppy, and is probably universal all over India.

Thus the differences between the root systems of varieties have been clearly shown and the economic significance illustrated.

Vegetable growing is an important phase of agriculture and one that is increasing at a rapid rate. In the United States practically every kind of soil is used for growing vegetables. Among the dozens or sometimes hundreds of varieties of the various vegetable crops, some undoubtedly have root systems more suited to a given soil environment than others. If the highest possible production is desired, attention should be given to selecting the variety that is not only climatically adapted but also best fitted to a particular soil, modified as favourably as lies within the power of the grower. It seems certain that a large part of the success in connection with the improvement of varieties consists in better adapting the absorbing system to the soil. As stated by one who has extensively worked on plant selection, breeding, root investigations, and other phases of agronomic science, "The more I learn about cultivated plants the more I am convinced that the future lies in the root-soils relation and in matters which influence it."

An intimate knowledge of the habits of growth of the root systems of vegetable crops will enable the grower to space plants to better advantage. It should also permit him to intercrop or grow in succession such crops or mixtures that the soil volume will have a better distribution of roots and thus permit of methods of more intensive cultivation. Similarly, by means of proper crop rotations and occasionally cultivating very deeply rooting crops, the subsoil may be kept in good condition and the effects of drought mitigated.

Interrelations of Plant, Soil, and Climate

In considering the importance of root relations in crop production, it should be clearly kept in mind that the plant, the soil, and the climate form a closely interlocking system of which no part should be overlooked or overemphasized. It is now rather generally recognized that climate and vegetation are the most important factors determining the character of the mature soil. 148 "The features assumed by the soil in its development from infancy, through youth, maturity, and old age, vary with the environment, especially with the climate and the vegetation." The effect of both climate and soil on the growth of above ground plant parts has long been known. It has only recently been clearly demonstrated that the environmental factors which affect the root are not only those of the soil immediately about it but also those affecting the shoot which is rightly a part of the complex. Through the shoot the root system is influenced by the aerial environment. The amount of light or the degree of humidity, temperature, etc. and the effect of these upon food manufacture, water loss, and other activities affect root development. In fact there is a rather close correlation between shoot and root development. Whatever affects the above ground growth of plants whether favourably or unfavourably is, in turn, very likely to exert an influence upon root development.

The complex relationship of plant, soil, and climate may be further illustrated in the use of fertilizers. They modify the habit of growth as well as the composition of the plant. For example, phosphates, when applied to soil upon which wheat or certain root crops are grown, promote deeper root penetration. This results in a greater water and nutrient supply for the plant. Earlier development and ripening may be promoted or drought mitigated. Soils thus fertilized will produce crops under an environment perhaps otherwise quite unfavourable. Vegetable production should be studied from the point of view of how roots and shoots of plants grow, and use should be made of the plant itself for indicating the direction of future research.

An adequate discussion of the environment of roots (the soil), how roots are built to perform their work, and root habits in relation to crop production has been so recently given in "Root Development of Field Crops" that further statement seems unnecessary. For a general discussion of the effects of irrigation, drainage, water content, aeration, temperature, nutrients, tillage practice, plant disease, and related phenomena upon root habit and their significance in crop production, the reader is referred to the same volume.

Activities of Roots in Subsoil—The great extent of the root systems of most vegetable crops and their usual thorough occupancy of the subsoil may at once arouse interest concerning the importance of the deeper soil layers. Experiments have shown that the roots of crops are active in the absorption of both water and nutrients even to the maximum depth of penetration. 174,165 Nutrients, when available, are taken from the deeper soils in considerable quantities, although to a lesser extent than from the soil nearer the surface which the roots occupy first and, consequently, at least in annual crops, where they absorb for the longest time. The deeper portions of the root system are often particularly active as the crop approaches maturity. Nutrients absorbed by them may produce a pronounced effect both upon the quantity and the quality of the crop yield.

Method of Root Study—In the present studies the direct method of root examination has been employed. It has been used by the writer and his coworkers in the excavation of hundred of root systems during the past 14 years and has proved very satisfactory. By the side of the plants to, be examined, a long trench is dug to a depth of about 5 feet and of convenient width. This affords an open face into which one may dig with hand pick and ice pick and thus uncover and make a careful examination of the entire root system. This apparently simple process, however, requires much practice, not a little patience, and wide experience with soil structure. In every case several plants were examined at each stage of development to insure an adequate idea of the general root habit. As the work of excavation progressed, the trench was deepened, if necessary, so that finally the soil underlying the deepest roots was removed. Frequently, the trenches reached depths of 6 to 11 feet.

Upon excavating the roots, detailed notes and careful measurements were made in the field. After several plants were examined, these notes were studied and any point that remained indefinite was at once clarified by further study. This method leads to a high degree of accuracy. Drawings of the root systems were made in the field on a large drawing sheet with pencil and later retraced with India ink. They were made simultaneously with the excavation of the roots and always by exact measurements. In the drawings the roots are arranged as nearly as possible in a vertical plane; that is, each root is placed in its natural position with reference to the surface of the soil and a vertical line from the base of the plant. In some cases the drawings represent the roots in their natural position in the surface foot of soil. In every case it was sought to illustrate the average condition of root development rather than the extreme. Although the drawings were made on a large scale, the rootlets were often so abundant that it was quite impossible to show the total number. Such drawings, however, carefully executed, represent the extent, position, and minute branching of the root system even more accurately than a photograph, for under the most favourable conditions, especially with extensive root systems, the photograph is always made at the expense of detail, many of the finer branches and root ends being obscured.

Conditions for Growth at Lincoln, Neb

Since roots, like above ground parts of plants, are greatly modified by environment, a brief statement will be given of the conditions under which the crops were grown.

Soils—The vegetable crops at Lincoln were grown in a fertile, dark-brown, Carrington silt loam. Preceding them white clover had been raised for 2 years. This was followed by a crop of maize. The soil was not only in an excellent condition as regards tilth but also was well supplied with humus. In general, it may be said that the soils of eastern Nebraska have a sufficient supply of all the essential nutrients to insure good crop yields, except that, owing to the systems of cropping, there may be a deficiency in available nitrates. The area was only slightly rolling, the general slope being toward the south. As usual, some differences in soil texture were found in excavating the root systems in different parts of the area. The following description of the soil profile represents an average condition.

The surface 12 to 14 inches was a mellow, very dark silt loam from which the roots were readily removed. At greater depths it intergraded into a dark-coloured, clay subsoil. This became quite sticky when wet and hard when dry. It exhibited a columnar or jointed structure, especially below 18 inches, cracking badly when it shrank upon drying. Roots were removed with much more difficulty from this soil layer. At about 3 feet in depth the subsoil became much lighter yellow in colour and graded rather abruptly into an easily worked, quite mellow, friable soil type approaching loess in many of its physical properties. This extended beyond the depth of greatest root penetration, i.e., over 12 feet. It was characterized by rusty streaks and numerous small calcareous areas and concretions. Earthworm burrows penetrated the soil to depths of 8 feet or more and numerous old-root channels extended almost as deeply. In many places extensive surface cracks reached depths of 4 to 5 feet. They were sometimes 1 inch in width. These crevices had been formed during past years of drought and were filled with rich, black, surface soil which had been washed or blown into them. Cultural operations in a large measure had also undoubtedly contributed to the filling.

Number and Size of Plats—The crops, in nearly all cases, were grown in triplicate plats so that early, midsummer, and later examinations could be made without disturbing the areas in which the plants were to make further growth. The plats, although somewhat variable in size, were in all cases large enough to permit of normal field development, each group of plants being entirely surrounded by plants of its kind. For example, the plats of lettuce, radishes, and onions were 12 feet long and 10 feet wide, the plants in the central portion of each plat being used for a single root study. In the case of the tomato, cabbage, eggplant, and other crops where each plant occupied considerable space, the plats were proportionately larger so that the plants examined were grown under the ordinary competitive conditions and mutual environment of garden crops.

Tillage—The field was plowed 8 inches deep early in the spring after all cornstalks, stubble, and dried weeds had been carefully removed. This was followed by repeated harrowing until an excellent, firmly compacted, moist seed bed of good soil structure was formed. On areas not immediately planted, weeds were removed by raking and hoeing. After the seeds were sown or the seedlings transplanted, further cultivation was done very shallowly with rake or hoe so as not to disturb the roots. Cultivation and weeding were repeated, as in ordinary gardening, when needed to prevent the growth of weeds and to keep the surface soil in a good condition of tilth. At no time did cultivation extend to greater depths than 1 inch.

Precipitation—The rainfall of 19.2 inches during the 6 months of the growing season was 2.3 inches below the normal, with periods of moderate drought occurring in May and again in July.

April with a deficiency of 1.2 inches had three well-distributed showers. In May efficient rains occurred on the eighth and fifteenth, with a monthly deficiency of 3 inches. June was a wet month with a total of 6.6 inches of precipitation, mostly in seven well-distributed rains. This was 2.3 inches above the mean. July had four rains during the latter half of the month and a total precipitation 1.8 inches below the mean. Five well-distributed showers fell in August, an excess of

1.2 over the mean. September had only 0.2 inch above normal, the rains also being well distributed.

As a whole the growing season was typical for eastern Nebraska. The spring was cool, frost occurring until about May 12.

Soil Moisture-It is a well-established fact that rainfall is only a very general indicator of soil moisture, since many other factors both climatic and edaphic intervene between precipitation and water available for plant growth. Hence, a study of the soil moisture in several of the plats was made from time to time throughout the growing season. Ideally this should have included determinations for each kind of crop at frequent intervals. Owing to the laborious task of excavating roots this was not attempted.

Temperature—The temperature of the air is an important factor not only directly in promoting metabolic processes but especially in connection with root studies in modifying humidity and affecting transpiration. The amount of water lost from the above ground parts reflects itself in the development and extent of the absorbing organs.

The soil above 6 inches was not only much warmer at times but also subject to greater temperature fluctuations than at greater depths. Thus roots lying close to the surface of the soil were subject to the influence of an environment which was quite different from that affecting those growing more deeply. It seems probable that surface-soil temperatures never became so high as to be injurious to root growth but that the absence of roots or their death in the shallowest soil was due to lack of a sufficient supply of moisture. 15,174 Soil temperatures are also of great importance because of their relation to disease-producing organisms that may greatly limit successful crop production.

Humidity—The loss of water through transpiration is directly controlled in a large measure by the amount of moisture already in the air. This is also an important factor governing the evaporation of water directly from the surface soil. Consequently, a record of relative humidity was obtained by means of a hygrograph appropriately sheltered with the recording apparatus about 5 inches above the soil surface

Evaporation—Evaporation was measured by Livingston's standardized, non-absorbing, white, cylindrical atmometers. These were placed in a bare area in the field with the evaporating surface at a height of 2 to 4 inches above the surface of the soil.

Conditions for Growth at Norman, Okla

The greater portion of root studies in Oklahoma was made during 1926. Some work was also done the preceding growing season and the root growth of several crops was followed during the intervening winter. Where root excavations were made in Oklahoma this is so stated in the text.

Soil—The crops were grown in a fine sandy loam soil that had been heavily manured for a number of years and used for growing vegetables. It was in excellent condition as regards tilth and amount of humus. The area was level and the soil quite uniform throughout. The surface foot had a reddish-brown colour, was quite sandy in texture but rather compact, and contained enough clay to exhibit a tendency to bake when it dried after heavy rains. In the second foot it was a light-chocolate brown and contained more clay. When dry, it became quite hard, especially at a depth of 18 to 24 inches.

The greater sand and smaller clay content of the soil prevented it from cracking and fissuring as did the soil at Lincoln. This, as will be shown, had a marked effect upon general root habit and branching. Roots penetrated the 18- to 24-inch soil layer with some difficulty, especially when it was rather dry. This was shown by their tortuous courses. The third foot consisted of a reddish-yellow sandy clay. It contained numerous, small, black concretions of iron 1 to 2 millimetres in diameter. After long periods of drought it became very hard. The next 18 inches were much sandier and had more and larger concretions but showed no change in colour. At greater depths the soil became still sandier, was mottled with reddish-brown spots, and contained many concretions. At 7 to 8 feet it was very compact.

Method of Planting and Tillage—The field was plowed to a depth of 8 inches late in March. Repeated disking and harrowing resulted in a good, compact seed bed. All of the crops were planted by hand, each in four to eight rows, 3.5 feet apart in order to permit of cultivation with a horse-drawn harrow. Shallow cultivation was given after each rain to prevent the formation of a surface crust and also to kill weed seedlings. A mulching fork, supplemented by a hoe when necessary, was used near the plants. In this manner a surface mulch 1 to 1.5 inches deep was maintained but in no case was a cultivation deeper. A good supply of moisture was present just below the mulch during almost the entire growing season.

Precipitation—The rainfall of 11.5 inches during the period when the crops were grown (April to July, inclusive) was 3.6 inches below the mean.

April with 1.9 inches had a deficiency of 1.3 inches and no efficient rainfall after the tenth. May had 2.3 inches which was somewhat (3.1 inches) below the normal amount. No efficient rain fell between May 8 and 30. During June precipitation was 1.6 inches less than the mean.

A precipitation of 2.2 inches, however, was fairly well distributed throughout the month. July had 5.1 inches rainfall, 2.4 inches above the mean, little moisture falling after the thirteenth.

Soil Moisture—Notwithstanding the rather limited rainfall, a good supply of water was present throughout the entire growing season. The fine sandy loam soil of the level field, loose from the cultivation of the previous season, had been thoroughly wet by the fall and winter rains. Water had entered the deeper subsoil and moistened it beyond the depth of greatest root penetration. Most of this water was retained in the spring and supplemented by rains in March. Thus, although the summer rainfall was light, a good supply of moisture was present throughout the entire growing season.

Other Factors—Another factor favourable to moisture conservation was a very cool, late spring. Warm weather did not begin until the second week in April. The surface soil warmed rapidly, there was enough moisture to promote prompt germination and vigorous growth, and the crops made excellent progress even under a light precipitation. The usual drought and high temperatures of July were replaced by a relatively cool, moist midsummer during which the plants made a continuous growth. The average day temperatures (6 a.m. to 6 p.m., inclusive) and the average daily temperatures (24 hours), obtained by means of a thermograph placed in the field.

Conditions for Growth during the Fall and Winter—The root development of certain crops was studied during the preceding fall and winter.' Following a dry, late summer in 1925, September was warm and had an abundance of moisture (5.4 inches).

During this period many long-lived vegetable crops showed a vigorous, renewed growth. Growth was continued, especially by the hardier plants, during an unusually cool October. Owing to favourable soil moisture and rather uniformly moderate temperatures during November and December, growth in several crops slowly continued. Following the cessation of growth in January, renewal of the development of both roots and tops was resumed during February which had a temperature of 8.2°F. above the mean.

Introduction Organic Vegetable Growing

Soil and Site

Soil Testing: Soil texture and structure. The hand test will tell you which of the soil types you have, Sandy Loam, Medium loam, Heavy Loam. Although Soil Texture cannot be easily changed, knowing the make up of your soil will give indications on how to make the best use of what you have with cultivations and fertility management. In general heavy soils will hold onto nutrients better but will tend to water log in wet seasons. Lighter soils will be hungry well drained and will heat up quickly in spring. Both extremes will benefit from the addition of humus

Soil Nutrients:, The test for NPK will inform decisions on what methods of fertility management are required and how best to plan the rotation.

pH and lime Test for acidity and recommend lime use. pH has a big influence on what nutrients are made available to the plant and on other soil fauna which are part of a healthy soil. In general terms alkaline soils are better for most vegetable crops whereas acid conditions give opportunities for ornamental plants. Rate of lime application can be calculated using the pH test results.

Site: Choice of site for vegetable patch usually limited,where possible avoid shade, water-logging and overhanging trees and tree roots. In general terms anything that will interfere with access to sun rain and nutrients.

Rotation and Planning

Rotation: Using a rotation of crops means that the same crop is not grown on a plot of land for two years running. Crops are divided into groups of related plants or crops requiring the same conditions and are rotated around pre determined plots within the growing area, so providing the opportunity to apply manures, lime and pest control measures according to the crops requirements. For example by keeping brassicas and potatoes on separate plots allows lime to be applied to brassicas to discourage club root while avoiding lime on potatoes which may encourage scab. The movement of crops around the growing area limits the spread of some soil borne diseases although this is less effective on smaller plots.

The rotation can be changed according to need. A traditional rotation would be 4 or 5 years say: Alliums followed by Brassica, followed by Roots followed by Legumes. If Green Manures are used extra years will be added to the rotation. The rotation will depend on what quantities of each crop are to be grown and whether a green manure crop is to be grown or if fertility is to be added using FYM or compost.

Alliums: Have need the highest nutrition due to primitive roots also respond to high nutrient levels. Must have high light levels. It is advisable to plant Alliums immediately after green manure or FYM if used

Brassica and Lime

Brassicas respond well to lime and as it helps prevent the establishment of club root. Lime can therefore be applied if needed before planting. Brassicas can follow Alliums which have had an input from green manure or compost so there is residual fertility from the green manure and any lime applied will help to stimulate microbial activity so liberating more nutrients.

Legumes

Legumes respond well to alkali soils and need reasonable nutrient levels to establish, thereafter they produce some nitrogen from their nitrogen fixing root nodules. Legumes can be flexible where they fit in a rotation as they have less soil associated problems.

Beets and Salads

These can fit with legumes or alliums, they respond well to high nutrient levels and do not have many soil borne problems.

Roots: Potatoes, parsnips and carrots, used late in the rotation so FYM and lime are broken down in previous plantings, so preventing forking of carrot or parsnip roots. Lime avoided for the roots part of the rotation as high lime content in the soil can increase incidence of common scab in potatoes. Compost or FYM can be added for potatoes but should be avoided for carrots and parsnips. The quantity of potatoes grown will determine if they should have a rotation of their own or if they are included with roots or possiblly legumes.

Green Manure: Green manures will replace nutrients used by crop plants if legumes such as sweet or red clovers are used. They can be used to prevent nutrient loss through leaching when non legumes are used notably over winter. They also provide humus and encourage soil fauna activity. If Green manures are to be used exclusively to replace nutrients and humus they will need to be grown ay least one in three years or more depending on the natural fertility of the soil and density of planting.

Compost

Making a compost heap: Garden style, square up to 2m square by about 2m high, this size ensures that oxygen can reach the inside of the heap while still being big enough to retain the heat needed for efficient working.

There are four factors in making a successful compost heap;

1. Oxygen, make the compost no more than 2m wide so air can pass into the centre of the heap, also the heap should not be over compressed so to prevent the access of air to the centre.

2. Water, necessary for the bacteria to develop and to allow the heap to heat up. When high nitrogen is added to a heap the composting process can be so rapid that it uses up all the available water and stalls the composting process. In winter however the heap must be protected from too much water as this will push the oxygen out causing anaerobic conditions which will also stall the process. Water may have to be added to the heap in hot dry conditions.
3. eat, this will be generated by the composting process but the design of the heap will help to retain it and kep the process going. The heap should be compacted so each layer is in contact with the next but so squashed as it removes all the air. Avoid water logging and drying out.
4. Carbon to nitrogen ratio: The nitrogen (leafy material) speeds up the process while Carbon (wood/straw lignified material) provides the material to turn in humus and slows the process. With too much N the compost will heat up quickly then stall. With too much C the compost will be cold and slow to start.

In the case of a stalled heap it can be restarted by turning the heap adding water, lignified (woody) material, leafy material as needed to get a proper balance.

Using Compost

Plant Cycle: The normal cycle for a vegetable plant is as follows;

1. Seed sown either direct in growing area, in module trays, seed trays or in see beds for later transplanting.
2. Transplanting were applicable either bare root or a module into final cropping area. All transplants are watered after planting
3. Weeding watering and staking and protecting from pest and disease as needed
4. Harvesting.

Seed Sowing and Transplant Production

Methods of sowing include manual placing seeds either into modules or direct into drills (braodcast sowing used for some spp). There are push seed sowers which will sow and cover a drill much quicker than by hand. Generally though most seed sowing on a garden scale will be manually placing the seed in the drill or tray. Some still sow seed into seed trays then prick of seedlings into pots or larger trays but with the advent of module trays this practice is seen less.

Pest and Disease Control

Rotation: Avoid diseases especially white rot onions, nematodes in potatoes and club root brassicas.

Beneficials: There are various beneficial organisms that will control pests some can be purchased and released deliberately. This method is most useful in an enclosed area like a greenhouse.

Knowledge of which predators are in your area will give the opportunity to encourage them with cultural methods. e.g. Beetle banks for ground beetles.

Crop covers: Effective against most larger pests including Rodents, Cabbage White and aphids.

Fleece, Mesh and Polythene are the Three main Pptions

Organic sprays: There are various materials which can be used under an organic system, these include soft soap, Ferramol for slugs and copper for potato blight. Although the organic grower would have to apply for a derogation for the use of some of these.

Cultivation Methods, Weed Control and Tools

Site clearance: If there is a significant covering of weeds these should be first be mown off and either collected then the soil inverted to bury annual weeds. Perennial weeds and roots should be removed by hand before inverting the top soil.

Alternatively a heavy gauge polythene or woven plastic sheet can be laid over the area to exclude light and kill weed growth. This can take up to 2 years if there is a heavy load of perennial weed.

Surface cultivation: Regular cultivation used prior to planting or sowing. Generally it is advisable to cultivate a shallowly as possible to avoid damaging beneficial soil organisms.

Hand tools: single digging with spades, forking over the surface

Mechanical methods: using rotary cultivators, plough.

Sub surface cultivation: Hand tools double digging, Bastard trench which can be used as initial site clearance or to break up a pan.

Mechanical methods using deep tines or sub soiling prior to rotary cultivating or shallow ploughing.

Weed Control after Initial Establishment

Soil covering using polythene or woven plastic membrane prior to sowing or planting: Using this method the soil covering sheet is removed prior to planting or sowing.

The objective is to exclude light so preventing germinated weed seedlings from growing. If the ground is covered for six weeks in early spring this can give four weeks before weed seedlings start to grow.

Weed control Mulches. bio degradable and reusable. Laid by machine hand tools or by hand. The crop plants are planted through the mulch so significantly reducing competition from weeds and incidentally reducing soil moisture loss during a dry summer. Stale seedbed: A method which involves allowing weeds to germinate to seedling leaf stage then hoeing off all the weed leaving a clean bed. The concept is to encourage germination of weed seed in the top layer of soil then hoe off to give a weed free plot for about 2 to 3 weeks.

Weed Control Tools Mechanical

If weeds are being controlled using mechanical methods, hoeing should be done before weeds are established.

Wheel hoe: Hoes dutch, draw and push pull hoes.

- Planting tools.
- Dibber, hand trowels
- Seed sowing tools.
- Seed sower push
- Wolf sower.

Protective Structures

Artificial Lighting/Growing Room: Can be useful early season transplant production to give reliable light and heat when its not available outside. Blue light spectrum needed to produce seedlings.

Polythene tunnels: Larger the better, not normally heated for frost protection, used for soil grown crops, module production. A proper rotation should be used to prevent rotation needed to prevent deterioration of the soil.

Greenhouse: Can be kept frost free, used to raise modules, also useful for growing tomatoes, as due to dry conditions blight is reduced.

Frames/raised beds: Useful for soil grown crops and can be used to bring on transplants. There is the option of changing the cover: polythene to increase temperature, mesh to keep out pests or shade to reduce temperatures.

Loose crop covers fleece and Mesh: Theses are laid directly over the crops, usually with no supporting structures. Fleece will raise the temperature of the soil and will prevent ingress by pests. They are however very light and can damage easily and get blown about letting in pests which then can breed unimpeded by predators.

Bibliography

Alec, William: *The Dairy Chemical Industry*, London: Longman Group Limited, 1971.

Avanish K. Tiwari: *Food Security and Global Economy*, Pentagon Press, Delhi, 2009.

Babita Bohra: *Dairy Farming in Mountain Areas*, Daya, Delhi, 2006.

Bhattacharya, Lata: *Biochemistry of Nutrition*, Discovery, Delhi, 2010.

Brij K. Taimni: *Food Security in 21 Century : Perspective and Vision*, Konark, Delhi, 2001.

Brock, H. : *History of Dairy Chemistry*, New York: Norton, 1992.

Chakraborty, Sudip : *Food Security and Child Labour : The Case of a Hazardous Occupation*, Deep and Deep, Delhi, 2011.

Chand, Ram: *Decision-Making of Dairy Beneficiaries: Role of Aspiration, Motivation and Knowledge*, Om Pub, Delhi, 2010.

Cohen, Lizabeth, *Making a New Deal, Industrial Workers in Chicago, 1919-1939* Cambridge University Press, 1991.

Collymore L.: *Fruit Production in Barbados*, Port of Spain, Trinidad and Tobago, 1996.

David, M.: *Ideas in Chemistry: A History of the Science,* New Brunswick, N.J.: Rutgers University Press, 1992.

De, Sukumar: *Outlines of Dairy Technology*, Oxford University Press, Delhi, 2001.

Dinesh Arora: *Biotech's Dictionary of Dairy Science*, Biotech Books, Delhi, 296.

Friberg, Stig. E..: *Food Emulsions*, New York: M. Dekker, 1997.

Guarti, Luigi, *The Valuation of Firms*, Blackwell Publishing, 1994.

Jacobson, M. : *Safe Food: Eating Wisely in a Risky World,* Washington, DC: Living Planet Press, 1991.

Jha, S N : *Dairy and Food Processing Plant Maintenance : Theory and Practice,* International Book Distributing, Delhi, 2006.

John Prince: *Dairy Farming: Being the Theory, Practice, and Methods of Dairying*, New York, 1888.

Kapoor, Ajay : *Dairy Science and Technology*, Vishvabharti Pub, Delhi, 2005.

Koli, P. A.: *Dairy Development in India: Challenges Before Co-Operatives*, Shruti Pub, Delhi, 2007.

Leena Parihar: *Dairy Microbiology*, Agrobios, Delhi, 2008.

Mohana Swamy : *Agro's Dictionary of Dairy Science,* Agro Botanical, 1995.

Mowery, D. C. : *Technology and Wealth of Nations*, Stanford: Stanford University Press, 1992.

Narang, R.K.: *Fruit and Vegetable Preservation Techniques*, APH Pub, Delhi, 2010.

Parihar, Pradeep and Leena Parihar: *Dairy Microbiology*, Agrobios, Delhi, 2008.

Patton, Stuart: *Principles of Dairy Chemistry*, Huntington, N.Y.: Krieger, 1976.

Pradeep Chaturvedi: *Food Security in South Asia*, Concept, Delhi, 2002.

Rao, M K : *Food and Dairy Microbiology*, Manglam Pub, Delhi, 2007.

Rao, P. Venkateshwara : *Dairy Farm Business Management*, Biotech Books, Delhi, 2008.

Saini, M.L. : *Plant Breeding and Crop Improvement*, CBS, Delhi, 1997.

Shepherd, D. : *Homeopathy for the First Aider,* Sussex, England: Health Science Press, 1953.

Shukla , Arvind N.: *Textbook of Dairy Chemistry,* Discovery Pub, Delhi 2010.

Singh, Gajendra : *Food for All : An Assessment of Food Security in Indian Context*, MD Pub, Delhi, 2007.

Singh, Vir and Babita Bohra: *Dairy Farming in Mountain Areas*, Daya, Delhi, 2006.

Singhal, K K and D D Sharma : *Advances in Dairy Animal Production,* International Book Distribut, 2003.

Sujata K. Dass: *Biotechnology and Food Security*, Isha Books, Delhi, 2004.

Thomson, Sutherland: *Grading Dairy Produce*, Medi World Press, Delhi, 1995.

Tomar, S.K. and Gunjan Goel : *Applied Dairy and Food Microbiology,* Agrotech, 2005.

Tyagi, Prasum: *A Textbook of Animal Physiology*, Dominant, Delhi, 2010.

Venkateshwara Rao: *Dairy Farm Business Management*, Biotech Books, Delhi, 2008.

Walstra, Pieter, and Robert Jenness: *Dairy Chemistry and Physics*, New York: Wiley, 1984.

William Alec : *The Dairy Chemical Industry*, London: Longman Group Limited, 1971.

Index

□□□